What's Your Paleo IQ?

The *Fossil News* Book of Paleo Quizzes, Puzzles & Brain Teasers

You're smart, but are you *fossil* smart?

WENDELL RICKETTS

Published by

Fossil News: The Journal of Avocational Paleontology

a project of FourCats Press

Published in the United States by FourCats Press
ISBN: 978-1-7348050-3-1
LCCN: 2022947965

FRONT COVER, CLOCKWISE FROM RIGHT 1) The right foot of the Moa, New Zealand, by Walter Mantell, Esq., from Gideon Algernon Mantell, *A Pictorial Atlas of Fossil Remains, Consisting of Coloured Illustrations Selected from [James] Parkinson's Organic Remains of a Former World,* and Edmund Tyrell Artis's *Antediluvian Phytology.* 2) Eurypterid by Ernst Haeckel; public domain. 3) *Ammonites obtusus,* an ammonite, from James Sowerby, *The Mineral Conchology of Great Britain; or Coloured Figures and Descriptions of Those Remains of Testaceous Animals or Shells, Which Have Been Preserved at Various Times and Depths in the Earth,* Vol. 2, 1812/1818. BACK COVER Left: *Platax macropterygius* from Louis Agassiz, *Recherches sul les Poissons Fossiles,* Vol. 4, 1833-1843. Right: *Cycladites nilsoni* from Grafen Kaspar Sternberg, *Versuch einer Geognostisch-Botanischen Darstellung der Flora der Vorwelt.* FACING PAGE *Pterodactylus longirostris* from Edward Pidgeon, *The Fossil Remains of the Animal Kingdom,* 1830. Respect and gratitude to the Biodiversity Heritage Library for putting these and thousands of other images at the disposal of the public.

What's Your Paleo IQ?

The *Fossil News* Book of Paleo Quizzes, Puzzles & Brain Teasers

Table of Contents

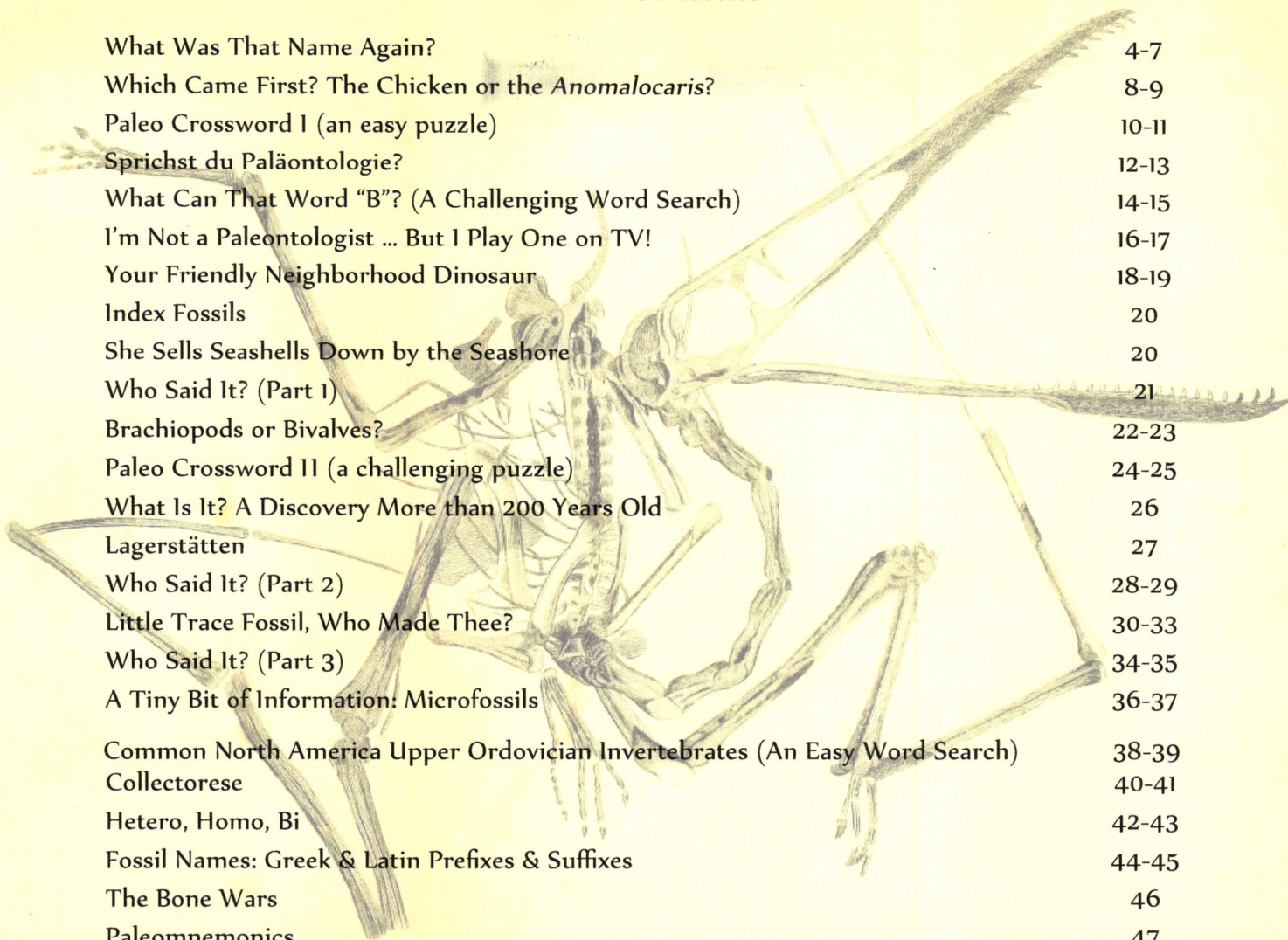

What Was That Name Again?

Species of every imaginable animal, plant, fungus, or protist have been named to give credit to thousands of scientists and discoverers, but quite a few have also been christened in honor of real-life musicians, movie stars, TV celebrities, and writers, both obscure and well known (and sometimes for reasons other than honor). Paleontologist Mats Eriksson, for instance, named a new Devonian marine worm (*Websteroprion*) for famed metalhead bass player, Alex Webster, of Cannibal-Corpse. Others have even been named for fictional characters.

A few folks seem to have earned more than their share of namesakes (Sir David Attenborough, the British broadcaster and natural historian, for example), and more famous names seem to crop up in a few groups of organisms more than others (trilobites probably win that prize). See if you can guess the famous person behind the name (most of them won't take long, but a few are puzzlers), and then try to imagine what kind of creature the name is attached to.

Torvosaurus gurneyi

Avalanchurus lennoni and *Avalanchurus starri*

Isbergia planifrons and *Warburgia crassa*

Bambiraptor

Darwinius masillae

Effigia okeeffeae

Chrichtonsaurus

Gagadon minimonstrum

Han solo

Scipionyx samniticus

Trierarchuncus

Jaggermeryx

Jenghizkhan bataar

Kootenichela deppi

Mackenziurus johnnyi

Leninia

Australopithecus afarensis

Megalonyx jeffersonii

Montypythonoides

Obamadon gracillis

Electrotettix attenboroughi

Amaurotoma zappa

Sauroniops

Mesoparapylocheles michaeljacksoni

Confuciusornis sanctus

Anisonchus cophater

Xenokeryx amidalae

Torvosaurus gurneyi: Fans of the Dinotopia book series may recognize this homage to illustrator and paleoartist, James Gurney, whose name is forever attached to this Late Jurassic megalosaur.

Avalanchurus lennoni and ***Avalanchurus starri***: These Silurian phacopid trilobites are named after two of The Beatles, John Lennon and Ringo Starr. Two other species of ***Avalanchurus*** are named for the 1960s musical duo, Simon & Garfunkle: *A. simoni* and *A. garfunkeli*. Cousins of ***Avalanchurus***, ***Struszia harrisoni*** and ***Struszia mccartneyi***, are named for the other two Beatles: Paul McCartney and George Harrison.

Isbergia planifrons and ***Warburgia crassa***: As the story goes, Swedish paleontologists Elsa Warburg and Orvar Isberg chose these less than flattering species names with each other in mind: Warburg christened *I. planifrons,* an Ordovician trilobite, with a species name that means "flat-headed"—or stupid—in Swedish. Isberg gave the name *W. crassa* to an Ordovician bivalve, knowing that "crassa" meant "thick"—as in fat.

Bambiraptor: A Late Cretaceous, bird-like dromaeosaurid dinosaur that was named after the Disney deer because of its small size.

Darwinius masillae: ***Darwinius*** is a basal, lemur-like primate from the Middle Eocene, and its genus name honors Charles Darwin on the occasion of the bicentenary of his birth.

Effigia okeeffeae: *E. okeeffeae* was an Upper Triassic archosaur that lived in what is now New Mexico. It was named after the painter Georgia O'Keeffe, who spent many years at Ghost Ranch, near where the specimen was found.

Crichtonsaurus: ***Crichtonsaurus*** was an ankylosaur that lived in China during the Late Cretaceous, and its name honors Michael Crichton, the author of *Jurassic Park* (most of the dinosaurs in *Jurassic Park* are Cretaceous species). ***Cedrorestes crichtoni***, an Early Cretaceous iguanodon, is also named for the author, and another Chinese ankylosaur was named with *Jurassic Park* in mind: ***Tianchisaurus nedegoapeferima***, whose species name is formed from the surnames of the film's stars: Sam NEill, Laura DErn, Jeff GOldblum, Richard Attenborough, Bob PEck, Martin FERrero, ArIana Richards, and Joseph MAzzello.

Darwinius masillae, reconstruction © Nobu Tamura, Licensed under CC BY 3.0.

Gagadon minimonstrum: The "Lady Gaga-toothed little monster" was described in 2014 on the basis of a fossil jaw and some unusual associated teeth. ***Gagadon*** was a hoofed mammal that lived in what is now Wyoming.

Han solo: When the description of this small agnostid Middle Ordovician trilobite from China was published in 2004, its author said it was named "Han" for the Han people of China and "solo" because it was the only member of its genus. Later, though, he admitted that his friends had dared him to name a species after a *Star Wars* character.

Scipionyx samniticus: ***Scipionyx***, known from a single specimen found north of Naples in 1981, was the first dinosaur known from Italy and is named for two people: Scipione Breislak, an eighteenth-century Italian geologist, and Scipio Africanus, the third-century BCE Roman military strategist and general who defeated Hannibal.

Montypythonoides: It's a shame this genus name for an extinct python didn't stick (it's now known as **Morelia**), because naming it for the comedy troupe Monty Python was just too perfect!

Obamadon gracillis: A small, extinct insect-eating Late Cretaceous lizard known from the Hell Creek and Lance Formations, was named for the forty-fourth president of the United States, Barack Obama. He also had an Ediacaran fossil, **Obamus coronatus**, named for him in 2018 because of the similarity of the creature's shape to Obama's ears.

Electrotettix attenboroughi: Found in amber from the Dominican Republic, this extinct pygmy locust is only one of the fossil (and other) species named for Sir David Attenborough. He also inspired the names for **Materpiscis attenboroughi**, a Late Devonian armored fish from Australia; **Mesosticta davidattenboroughi**, a Cretaceous damselfly; **Microleo attenboroughi**, a miniature marsupial lion from the Early Miocene; and **Attenborosaurus**, an Early Jurassic plesiosaur from England.

Amaurotoma zappa: This Permian gastropod from Nevada was named for Frank Zappa (1940-1993), the genre-bending American musician, composer, and musical satirist. Zappa also lent his name to an Early Miocene gerbil-like rodent, **Vallaris zappai**.

Jefferson's Chesapeake Bay Pecten. Left valve of Chesapecten jeffersonius (Say, 1824). Yorktown Formation, Lower Pliocene, Williamsburg, VA. Photo: James St. John. Licensed under CC BY 2.0.

9

Sauroniops: The genus name of this carnivorous carcharodontosaurid dinosaur from the Late Cretaceous of Morocco means "The Eye of Sauron" and is, of course, a reference to the terrible mountain entity from Tolkien's *Lord of the Rings*.

Mesoparapylocheles michaeljacksoni: This extinct Cretaceous hermit crab species was named for pop singer Michael Jackson because paleontologists discovered it on the same day they learned of Michael Jackson's death in 2009.

Confuciusornis sanctus: This early bird relative from the Early Cretaceous of China, and an ornithischian dinosaur, **Tianyulong confuciusi** from the Early Cretaceous Jehol group in Western Liaoning Province, China, were both named in honor of the Chinese philosopher, Confucius, who lived between 551 BCE and 479 BCE.

Anisonchus cophater: This story comes from the life of Edward Drinker Cope published in 1931 by American paleontologist, geologist, and eugenicist, Henry Fairfield Osborn. In it, Osborn quoted a letter from Cope in which Cope explained that he had given the placental Paleocene mammal, **Anisonchus**, the specific name "*cophater*" for "all the Cope-haters who surround me." Marsh, in his turn, named a marine reptile after Marsh: **Mosasaurus copeanus**. A hundred years later, an American evolutionary biologist had the final word: he named yet another primitive Paleocene ungulate **Oxyacodon marshater**.

Xenokeryx amidalae: Fossils of this ancient ancestor of the giraffe from Spain were first described in 2015. The leader of the team who discovered the new species said they chose the name "*amidalae*" in honor of the *Star Wars* character Padmé Amidala Naberrie because its horns resembled the elaborate hairstyles she wore as queen of Naboo.

Trierarchuncus: You might consider this one a bit of a cheat. ***Trierarchuncus*** was a small, long-legged, bird-like but flightless dinosaur. It had a single large claw on each hand and, thus, its name means "Captain Hook." ***Trierarchuncus*** isn't really named for the **Peter Pan** character, though. Its name comes from the combination of "*triarch*" (the captain of a trireme in classical Greece) and *uncus*, meaning "hook" in Latin. Its specific name, ***prairiensis***, "from the prairie," refers to the plains of eastern Montana where its fossils were first discovered.

Jaggermeryx: An extinct genus of semiaquatic anthracothere (ungulates related to hippopotamuses) from the Early Miocene of Egypt, ***Jaggermeryx*** was named for Rolling Stones front man, Mick Jagger, in part because it was believed to have large, fleshy lips. Jagger also has a trilobite to his credit (***Aegrotocatellus jaggeri***) and a tiny Permian marine gastropod (***Anomphalus jaggerius***).

Jenghizkhan bataar: During several years in which the classification of the tyrannosaurid Mongolian dinosaur, ***Tarbosaurus bataar***, was a matter of debate, the American writer and paleontologist, George Olshevsky, proposed the genus name ***Jenghizkhan*** in honor of Genghis Khan, the twelfth-century Emperor of the Mongol Empire. In fact, Olshevsky recognized three distinct genera of Late Cretaceous Mongolia "meat-osaurs," but, over the last twenty years or so, they have been revised into a single accepted genus: ***Tarbosaurus***.

Kootenichela deppi: This extinct arthropod, a distant ancestor of lobsters and scorpions, was named for actor Johnny Depp, but for an interesting reason: It has large claws with elongated spines that reminded the paleontologist who described it of Edward Scissorhands, the character Depp played in the 1990 movie of the same name.

Mackenziurus johnnyi: Another Silurian phacopid trilobite, ***Mackenziurus*** honors four members of the American punk band, The Ramones, active from 1974-1996. ***M. johnnyi*** was named for Johnny Ramone, ***M. joeyi*** for Joey Ramone, ***M. deedeei*** for Dee Dee Ramone, and ***M. ceejayi*** for C. J. Ramone

Leninia: An extinct Cretaceous ichthyosaur, ***Leninia*** was named for Vladimir Lenin, the head of Soviet Russia from 1917 to 1924 and of the Soviet Union from 1922 to 1924.

Australopithecus afarensis: A trick question. The actual scientific name, ***A. afarensis*** doesn't have anything to do with a famous individual, but you may recall the nickname by which this proto-human was known in the media after her discovery in 1974: "Lucy." And "Lucy," of course, is a reference to the Beatles' song *Lucy in the Sky with Diamonds*, which, according to the legend, was played often and loudly in the expedition camp in the Awash Valley, Ethiopia, where her fossils were found.

Kootenichela deppi

Megalonyx jeffersonii: A large, extinct Pleistocene ground sloth, named for the third president of the United States, Thomas Jefferson, who was an avid amateur paleontologist. French zoologist Anselme Desmarest christened ***M. jeffersonii*** in 1822 and, exactly 100 years later, Jefferson's name was used once more for the North American mammoth, ***Mammuthus jeffersonii*** (***M. jeffersonii*** is now largely considered a synonym for other ***Mammuthus*** species). Jefferson has at least one more fossil name to his credit: the large and striking bivalve, ***Chesapecten jeffersonius***, believed to be the first North American fossil to be described in official scientific literature (1687). This Pliocene mollusk is the state fossil of Virginia.

Which Came First? The Chicken or the *Anomalocaris*?

OK, that's too easy. But some of the ones below might stump you. Can you arrange these important geological and paleontological milestones in Earth's history in the proper chronological order?

____ First cockroaches

____ First horses

____ Ediacaran organisms appear

____ End-Permian extinction

____ First multicellular life

____ First insects

____ First sexual reproduction

____ First land plants

____ Western Interior Seaway opens

____ Great oxygenation event

____ *Tiktaalik*

____ First dinosaurs

____ First *Homo sapiens*

____ Pangaea forms

____ First primates

____ First sharks

____ First life appears

____ First amphibians

____ Cambrian Explosion

____ Mammals first appear

____ "Lucy" (*Australopithecus afarensis*)

____ Extinction of dinosaurs

____ First green algae

____ First reptiles

____ First monkeys

____ The Green River Formation begins

____ *Archaeopteryx*

____ First atmospheric oxygen

____ Extinction of *Otodus megalodon*

____ First ferns

____ First fish with vertebrae/true bones

1.	First life appears	~ 3.8 billion ya
2.	First atmospheric oxygen	~ 3.5 billion ya
3.	Great oxygenation event	~ 2.4 billion ya
4.	First sexual reproduction	~ 1.2 billion ya
5.	First green algae	~ 1 billion ya
6.	First multicellular life	~ 800 million ya
7.	Ediacaran organisms appear	~ 600 million ya
8.	Cambrian Explosion	~ 540 million ya
9.	First fish with vertebrae/true bones	~ 485 million ya
10.	First insects	~ 480 million ya
11.	First land plants	~ 470 million ya
12.	First sharks	~ 425 million ya
13.	*Tiktaalik*	~ 375 million ya
14.	First amphibians	~ 370 million ya
15.	First ferns	~ 360 million ya
16.	Pangaea forms	~ 335 million ya
17.	First cockroaches	~ 320 million ya
18.	First reptiles	~ 310 million ya
19.	End-Permian extinction	~ 252 million ya
20.	First dinosaurs	~ 240 million ya
21.	Mammals first appear	~ 225 million ya
22.	*Archaeopteryx*	~ 150 million ya
23.	Western Interior Seaway opens	~ 100 million ya
24.	Extinction of dinosaurs	~ 66 million ya
25.	First horses	~ 55 million ya
26.	The Green River Formation begins	~ 53.5 million ya
27.	First primates	~ 50 million ya
28.	First monkeys	~ 34 million ya
29.	Extinction of *Otodus megalodon*	~ 3.6 million ya
30.	"Lucy" (*Australopithecus afarensis*)	~ 3.2 million ya
31.	First *Homo sapiens*	~ 200,000 ya

6

A cast of the famous Tiktaalik roseae specimen, half-fish, half-amphibian, from the Devonian of Canada. Field Museum of Natural History, Chicago. Photo: James St. John. Licensed under CC BY 2.0.

Across

4. Along with 10 down, the famous Cretaceous exposure in Montana that is known not only for fossils of dinosaurs but of fish and plants as well.
6. In most (but not all) cases, the deeper a fossil is found, the _____ it is.
7. Paleozoic invertebrate whose name means "three lobed."
9. A counterpart also has one of these.
12. They're only found in Asia and Africa today, but these mammals evolved in North America.
14. The vast majority of fossils are found in this kind of rock.
17. Acknowledged as the first scientist to say that dinosaurs had feathers.
18. This dinosaur never existed but still starred in *Jurassic World*.
19. The "egg thief," though the name is no longer thought to be accurate.

Down

1. Perhaps the best known among the prehistoric animals found as fossils at the La Brea Tar Pits.
2. Its name means "quick plunderer" or "swift seizer."
3. Where there's fauna, look for this.
5. All these are in the Class Anthozoa, but not all anthozoans are these.
8. A long period of low temperatures that results in the formation or expansion of glaciers.
10. See 4 across.
11. These stomach-footed mollusks are commonly called this.
13. A barnacle and a honeybee are both this.
15. The Chicxulub crater is buried beneath the _____ Peninsula in Mexico.
16. David Schwimmer played one on TV.
20. Fossilized tree resin that sometimes contains the trapped remains of insects and other animals.

For further reading ....

19 Across: The "Oviraptoridae" page from the University of California Museum of Paleontology (ucmp.berkeley.edu/diapsids/saurischia/oviraptoridae.html) has great information about how the so-called "egg snatchers" really lived.

15 Down: Dr. Walter Alvarez, known for the theory that dinosaurs were killed by an asteroid impact at the end of the Cretaceous Period, which he and his father, Nobel Prize-winning physicist Luis Alvarez, developed together, recounted the story of that discovery in his book, *T. rex and the Crater of Doom (Princeton, NJ:* Princeton University Press, 1997). Novelist Douglas Preston's 2019 article, "The Day the Dinosaurs Died," is a fascinating account of more recent study of the KT (or KPg) Extinction (*The New Yorker*, April 8, 2019, https://www.newyorker.com/magazine/2019/04/08/the-day-the-dinosaurs-died).

20 Down: For more on the ethical problems inherent in the study of amber from Myanmar, and of "parachute science" in general, see Ortega, Rodrigo Pérez (2022, 29 September). Violent Conflict in Myanmar Linked to Boom in Fossil Amber Research, Study Claims. *Science.org* (www.science.org/content/article/fossils-burmese-amber-offer-exquisite-view-dinosaur-times-and-ethical-minefield); as well as the underlying study that Ortega discusses: Raja, N. B., et al. (2022, February). Colonial History and Global Economics Distort Our Understanding of Deep-Time Biodiversity. *Nature Ecology & Evolution, 6,* 145-154 (www.nature.com/articles/s41559-021-01608-8.pdf).

Paleo Crossword 1

... an easy puzzle

Sprichst du Paläontologie?

More than a few words in common use in geology and paleontology are derived from other languages, but a surprising number are either borrowed from or influenced by German. How many of the ones below do you know? And even if you recognize their meaning, what do you know about their origins?

auroch	K/Pg	plattenkalk
bauplan	Lagerstätte	steinkern
Darwinius masillae	Neanderthal	Ur

auroch: A borrowed word from early modern German. The auroch is an extinct species of large wild cattle (*Bos primigenius*) that inhabited Europe, Asia, and North Africa, and is the ancestor of today's domestic cattle. Neolithic humans apparently made at least a couple of attempts to domestic aurochs, which resulted in two lines of cattle that we know today: zebus (humped species, generally speaking) and taurines (most modern breeds). Genetic testing has shown that European bison, which were extinct in the wild by the end of the 1920s, shared DNA with the auroch. The last known auroch died in 1627 in Poland.

bauplan: Literally, "body plan." Bauplan is a common term used to refer to a set of morphological features common to all or most members of a phylum of animals. In general, the bauplan can be imagined as a sort of "blueprint" for how an animal body is made, and it includes things like symmetry; segmentation; and kind, number, and position of limbs, etc. All echinoderms, for example, share the same basic bauplan: five-fold radial symmetry and a water-vascular system. On top of that basic structure, animals as different as cystoids, crinoids, sea cucumbers, and sand dollars are "built."

Darwinius masillae: This is a bit of a cheat because *Darwinius masillae*, the name given to the extraordinary fossil of a primitive Eocene primate discovered in 1983, comes only tangentially from German. But it still counts because the specific name *masillae* derives from an old German term ("Masilla") for the Messel area in southwest Germany where the fossil was found: the world-famous "Messel Pit," which has yielded extraordinary fossils for years.

K/Pg: The K/Pg boundary (referred to in the past as the K/T) boundary) marks the end of the Cretaceous and the beginning of the Paleogene (or of the Tertiary) and, more specifically, is the demarcation of the great end-Cretaceous extinction. But why isn't it called the "C/Pg" event then? The "K" comes from the German word for Cretaceous, *Kreide* (the *Kreidezeit* is the Cretaceous period).

Lagerstätte (plural: Lagerstätten): This is probably a familiar term to anyone interested in fossils. Literally a "place of storage" in German, a Lagerstätte is a deposit in which extraordinary fossils with near-perfect preservation are found, sometimes including soft tissues, feathers, and other rarely fossilized remains. Paleontologists believe these formations represent anoxic (oxygen-depleted) environments in which few or no bacteria or other organisms could disturb a creature's remains before fossilization. Some of the best known Lagerstätten include the Burgess Shale, Mazon Creek, the Maotianshan Shales in China, and the Green River Formation.

Neanderthal: Neander Thal—literally Neander Valley—is the place in western Germany where bones of "Neanderthal man," a species of hominin that lived as early as 450,000 years ago, were first discovered in 1856. Experts remain divided as to whether Neanderthals were a separate species (*Homo neanderthalensis*) or a subspecies of modern humans (*Homo sapiens neanderthalensis*), but some of the debate hinges on the way the word "species" is defined.

plattenkalk: Literally, "limestone slab." Plattenkalks are very finely grained limestones laid down under conditions in which no organisms reworked or disturbed the sediment. The result is thinly bedded, finely laminated limestones in which exceptionally detailed fossils or imprints of organisms occur. The Late Jurassic Solnhofen Plattenkalk in Bavaria is one famous example.

steinkern: "Stone cores" are a stony mass that represent a specific kind of fossilization. Mud or sediment entered a hollow natural object (such as a bivalve shell), consolidated, and remained as a cast after the shell (or other object) was worn away. Steinkerns are sometimes very faithful molds of skulls, shells, and other organic remains.

Ur: Strictly speaking, "Ur" is used as a prefix rather than as a separate word, but in German it denotes something that is primitive, original, or earliest. *Archaeopteryx*, for example, is sometimes called the Ur-Vogel ("original bird" or "first bird"), and Panthalassa, the ancient sea that surrounded Pangaea, may be referred to as the Ur-ocean.

A M B J D N W J I K T P E Q Q G D G E E U A A T D N W J I S P T U G
Y B U A T B N A Y I O S U R U A S O L I S A B O P E X M E J C P R O
S P M F R A D N W J I N T C R W M O P W S A C D C D N W J T E P B A
B E L L E R O P H O N R Y K T P E O Q G D G O T T N W J I S Y P I I
W Y V M Z R D A A Q R N G F Q R N W M O U P B T A G L R C E Q F O M
H A M V F E U D K T P E Q F G D G U G B O I P M E D S F S V A C S I
R R V W F M L Z O C E P D M D V P D F I A L D N W D N W J L E P T A
G P W C J I C D B L F M V P E U E G H M L R I J T G I J T A C O R W
L G C B T A V W D M B E R D J V E C C W R A Y A F L F D Y V X E A O
R T B I Y N B P A R B J D E G I A A B H P D L O L V K E R I V Y T B
O U I E P D F Q B Y A G A J E R F D R R T Z D P N B C J B B K V I M
I Q E D A W Q P P E K E M G B L V Z A G K A U Y A Y C D A P R M G V
E M B W J I C E P D K P P E D N W J N L C I P M E D X E L D P V R W
E F K E E D V L A Y E B E L N W X H C R U N W J I C E F N E V W A C
I L R O L S A P I W R V V C N W J I H E P D F B W J I C Z J D C P B
E F Z D R E L F S E Q W P V D N W J I I P M E D I G S S D G E B H I
F B A B A B M V U Q W C R B R Y O Z O A A W J I S P F Q A N W M Y E
M L D T W R S N L Y G O M E L C B C P T X H I J U Z E C S F R E F D
P A R Y O E D T I C F I T P E O E P O E V R D C G U W D R G P D M C
D S X H I J U Q B T D E W J I V X H D J T G E K B A K U A V V N P D
E T A A E D V L A Y E D G C T N V A E D D L A Y K A P Y L L G D M U
J O D D W J I C E P E S V B A R K K B U P R G T L A W J I B I A E L
G I Z A W J I P B A B I E L I P B R O N T O S A U R U S D E T S J C
E D N W J I C E W A F P P F M V I E L I B H N A K U G X F Q R N M M
L K T P E Q Q G R G F U F M O M E L I D V Y Z O P W S A F R O T E B
C N O V E C I N A E A D D N W I P M D D E R D T N W J I C E P D L A
V A I B T H U C D B N W J I C V P L E V R D A D Y A W J L C D V C S
E X W D E M U W Z U G A P K T Q A N W J I X S P L E V K A X F Q V R
J C R E B N O V E N O V E C I Q D N W J I I P M E D E C S E E F Q I
P Q B R A W Q P P E B I O M I N E R A L I Z A T I O N T A E O Q G D
G R O Z T P E Q F G D P D G D V K U V L A V L V L L A Y B T E C D V
E W L D K E B I O T U R B A T I O N I V X H I J T G E O A A W J I W
N E P O P F M V P E U I F L A Y K T E N Y J I C L A P K T Q A W J I

What Can That Word "B"? (A Challenging Word Search)

The Word Search on the facing page features words in the paleosciences that begin with the letter "B." Below are the clues to the hidden words. Check the answer key if you get stuck.

1. Animals with hard "valves" on the upper and lower surfaces, much more common in the Paleozoic than today.

2. Mollusk whose shell is composed of two hinged parts.

3. Is it really *Apatosaurus*?

4. Terrestrial locomotion whereby an organism moves on its two rear limbs.

5. Mostly small, freshwater crustaceans that feed on plankton and detritus.

6. An extinct order of cephalopods with an internal skeleton.

7. "Moss animals."

8. The asteroid that ended the Cretaceous was one, generally speaking.

9. The process by which living organisms produce minerals, often to stiffen their tissues.

10. A genus of extinct Paleozoic marine gastropod.

11. In phylogenetics, a term for groups thought to possess ancestral characteristics.

12. "Body plan."

13. He advised on *Jurassic Park*.

14. Discoverer of *Tyrannosaurus*.

15. The science of assigning ages to rock strata through the fossils they contain.

16. The reworking of soil or sediment by animals or plants.

17. The collection of all organisms in a geographic region or time period.

18. What the Lower Cretaceous is named in the European system.

19. A large, predatory Eocene whale.

20. A giant, Early Cretaceous theropod with crocodile-like jaws.

21. An extinct stemmed echinoderm.

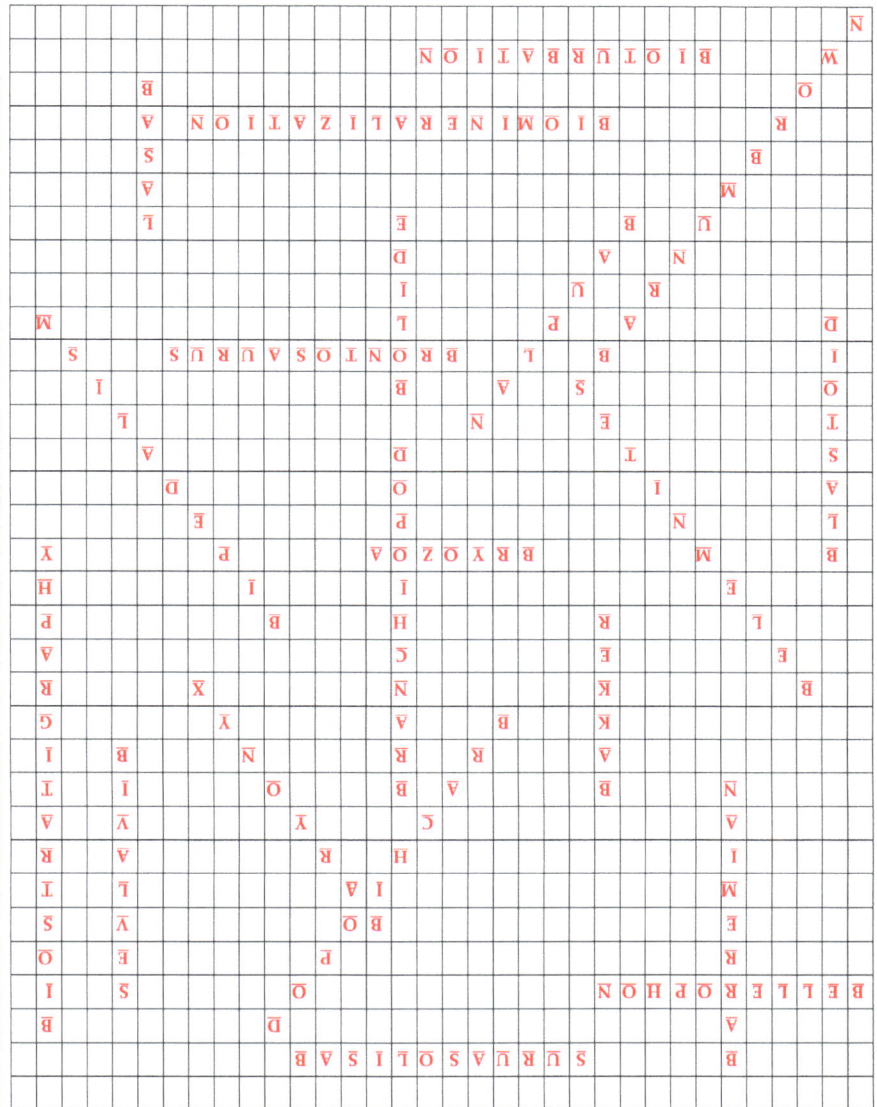

1. brachiopods
2. bivalves
3. *Brontosaurus*
4. bipedalism
5. branchiopod
6. belemnite
7. bryozoa
8. bolide
9. biomineralization
10. *Bellerophon*
11. basal
12. Bauplan
13. Bakker
14. Barnum Brown
15. biostratigraphy
16. bioturbation
17. biota
18. Barremian
19. *Basilosaurus*
20. *Baryonyx*
21. blastoid

I'm Not a Paleontologist ... But I Play One on TV!

5 points for each character you correctly match with the actor + 5 points if you name the movie, TV series, or episode.

1. David Schwimmer
2. Cary Grant
3. Laura Dern
4. Will Ferrell
5. Vera Farmiga
6. Treat Williams
7. Dean Cain
8. Sam Neill
9. Thomas F. Duffy

10. Kate Winslet
11. James Gandolfini
12. Geordie Johnson
13. Laurence Naismith
14. Edward Everett Horton
15. Vicki Lewis
16. Sally Hawkins
17. Henry Woronicz
18. Julianne Moore

19. Jack Black
20. Jason Statham
21. Mercedes Morán
22. B.D. Wong
23. Steve Carell
24. Jason David Frank
25. Jeff Bridges
26. Elizabeth Lackey
27. Bridget Fonda

Scores

240 - 270: Congratulations! You're a pop culture whiz!

175-235: Not bad, but don't try out for *Jeopardy*! just yet.

170 and below: You're spending way too much time away from the TV!

1. **David Schwimmer:** Ross Geller | *Friends* (1994-2004)

2. **Cary Grant:** David Huxley | *Bringing Up Baby* (1938)

3. **Laura Dern:** Dr. Ellie Sattler | *Jurassic Park* (1993), *Jurassic Park III* (2001), and *Jurassic World Dominion* (2022).

4. **Will Ferrell:** Dr. Rick Marshall | *Land of the Lost* (2009)

5. **Vera Farmiga:** Dr. Emma Russell | *Godzilla: King of the Monsters* (2019)

6. **Treat Williams:** Theodore Lytton | *Journey to the Center of the Earth* (TV miniseries; 1999)

7. **Dean Cain:** Dr. Robert Trenton | *New Alcatraz* (aka *Boa*) (2001)

8. **Sam Neill:** Alan Grant | *Jurassic Park* (1993), *Jurassic Park III* (2001), and *Jurassic World Dominion* (2022).

9. **Thomas F. Duffy:** Dr. Robert Burke | *The Lost World: Jurassic Park* (1997)

10. **Kate Winslet:** Mary Anning | *Ammonite* (2020)

11. **James Gandolfini:** Othniel Charles Marsh | *Bone Wars* (a cheap trick; this film has been in development at HBO since 2013 and no release date has been announced; see Steve Carrell.)

12. **Geordie Johnson:** Barclay Blake | "Dinosaur Fever" (*Murdoch Mysteries* 2:3) (2009)

13. **Laurence Naismith:** Professor Horace Bromley | *The Valley of Gwangi* (1969)

14. **Edward Everett Horton:** Alexander Lovett | *Lost Horizon* (1937)

15. **Vicki Lewis:** Dr. Elsie Chapman | *Godzilla* (1998)

16. **Sally Hawkins:** Dr. Vivienne Graham | *Godzilla* (2014) and *Godzilla: King of the Monsters* (2019)

17. **Henry Woronicz:** Professor Forra Gegen | "Distant Origin" (*Star Trek: Voyager* 3:23) (1997)

18. **Julianne Moore:** Dr. Sarah Harding | *The Lost World: Jurassic Park* (1997)

19. **Jack Black:** Sheldon Oberon | *Jumanji: The Next Level* (2019)

20. **Jason Statham:** Jonas Taylor | *The Meg* (2018)

21. **Mercedes Morán:** Carlota Schönfeld-Müller | *Las Rojas* (2021)

22. **B.D. Wong:** Dr. Henry Wu | *Jurassic Park* (1993) and *Jurassic World Dominion* (2022). (Technically, he's a geneticist, but we could call him a paleogeneticist.)

23. **Steve Carell:** Edward Drinker Cope | *Bone Wars* (a cheap trick; this film has been in development at HBO since 2013 and no release date has been announced; see James Gandolfini.

24. **Jason David Frank:** Tommy Oliver | *Power Rangers* (1993)

25. **Jeff Bridges:** Jack Prescott | *King Kong* (1976)

26. **Elizabeth Lackey:** Dr. Jessica Platt-Trenton | *New Alcatraz* (aka *Boa*) (2001)

27. **Bridget Fonda:** Kelly Scott | *Lake Placid* (1999)

Your Friendly Neighborhood Dinosaur

Based on the clues below, what's the dinosaur called?

1. Venda was already having trouble with the paparazzi after *Jurassic Park* made her famous, but then she got arrested for stealing!

2. Tamara hails from Mongolia, where she's considered alarming!

3. Gael looks like a chicken—until you see those giant claws.

4. Ira mostly likes to read his Arthur Conan Doyle novels, but whenever he goes out, people find him just plain irritating.

5. Psyche is a big parrot at heart; too bad crackers hadn't been invented yet.

6. Everyone agrees: Maya is a good mom.

7. Paresh's father often gets exasperated because his kid has such a thick skull.

8. Denequa hears people complaining all the time about being scratched by their pet cats and thinks, "They should see how terrible my claws are!"

9. Everybody likes Sten, but no one's eager to catch the tail end of him.

10. Ed's family might not be Mormons, but there sure are a lot of them in Montana!

11. Arjan is proud of his feathers; he got them in Germany.

12. Hania always takes her family out for roast duck on special occasions; if anyone wants more, she tells the waiter, "Just put it on the bill!"

13. Aphrodite has always followed a strict vegetarian diet, but she still weighs in at about twenty-two tons.

14. Panchali had to have a special motorcycle helmet designed to accommodate her fancy hair-do.

Top: *A reconstruction of* Archaeopteryx lithographica. *Image courtesy of the DataBase Center for Life Science. Licensed under CC BY 4.0. At right: Reconstruction of* Irritator, *based on the mounted skeleton at the National Museum of Rio de Janeiro. Image by Fred Wierum. Licensed under CC BY 3.0. Facing page: Parasaurolophus walkeri restoration by Steveoc86. Licensed under CC BY 2.5.*

1. *Velociraptor*: These turkey-sized predators from the Late Cretaceous lived in what is now Asia, and their most famous feature is a big, sickle-shaped claw on each hindfoot.

2. *Tarbosaurus*: Another Late Cretaceous dinosaur that also lived in Asia (Mongolia and China, mostly) and is a relative of other tyrannosaurid dinosaurs. Some paleontologists believe *Tarbosaurus* and *Tyrannosaurus* are the same creature but represent Asian and North American variants. The name literally means "alarming lizard."

3. *Gallimimus*: Named a "chicken mimic" for the shape and arrangement of its neck vertebrae, *Gallimimus* was a fast-running dinosaur that grew to about six feet in height and as much as twenty feet in length.

4. *Irritator challengeri*: A spinosaurid from the Early Cretaceous of what is now Brazil. The only known specimen, a nearly complete skull, had been so damaged by private collectors (before it found its way into a museum) that paleontologists were "irritated." The species name honors the character of Professor Challenger from Doyle's sci-fi novels.

5. *Psittacosaurus*: An Early Cretaceous, non-avian dinosaur whose most unusual feature was a powerful, parrot-like beak. *Psittacosaurus* was a ceratopsian, and its fossils are so common in Asia that it has become one of the best-known dinosaur genera.

6. *Maiasaurus*: The first fossils of *Maiasaurus* were found with nests and eggs, and paleontologists Jack Horner and Robert Makela named this Late Cretaceous dinosaur the "good mother."

7. *Pachycephalosaurus*: *Pachycephalosaurus* lived during the Late Cretaceous in what is now North America. Its thick skull roof formed a "dome" on its head that some believed was used in intra-species combat. Because only skull remains have been found, the animal's general anatomy is poorly known.

8. *Deinonychus*: An Early Cretaceous raptor known from the area that is now the western U.S. *Deinonychus*'s sleek body and long, thin tail were part of what led paleontologists in the 1960s to rethink the way they had conceived of dinosaurs up until then—some, like *Deinonychus*, were agile and active rather than slow-moving and ponderous.

9. *Stegosaurus*: Famous for its spiked "thagomizer," *Stegosaurus* was an armored dinosaur from the Late Jurassic. It was an herbivore and needed all that armor to fend off meat-osaurus predators.

10. *Edmontosaurus*: A large herbivorous hadrosaur and one of the last non-avian dinosaurs. In the exposures where it is found, *Edmontosaurus* is estimated to account for one-seventh of all the fossils recovered.

11. *Archaeopteryx*: The first complete specimen of this Late Jurassic "Ur-Vogel" or "primeval bird" was found in 1861. A small number of beautifully preserved specimens later came to light in the area around Solnhofen, Germany, cementing *Archaeopteryx* as an exciting transitional fossil between non-avian dinosaurs and birds.

12. *Hadrosaurus*: One of the "duck-billed dinosaurs," *Hadrosaurus* lived in North America during the Late Cretaceous. Its fossils are scarce, and only one partial specimen—from New Jersey, where it is the state fossil—is known.

13. *Apatosaurus*: Othniel Charles Marsh first described this Late Jurassic North American dinosaur. Its name means "deceptive lizard," which has turned out to be fitting given that debates have gone on for years about the similarities (or differences) between *Brontosaurus* and *Apatosaurus*.

14. *Parasaurolophus*: Another herbivorous hadrosaurid dinosaur of the North American Late Cretaceous. Its most striking feature is a hollow cranial crest. Some believe the crest played a role in temperature regulation, but others maintain, because the crest is connected through tubes to the nasal passages, that it was used to communicate through low-frequency sounds.

Index Fossils

An "index fossil" is:

a) Any specimen of a new species found in the same type locality.

b) A common, broadly-distributed species that can help pinpoint the age of an exposure.

c) A fossil whose appearance marks the beginning or the end of a mass extinction.

d) The specimen on whose basis a new genus, species, or other taxonomic group is described for the first time.

b) A common, broadly distributed species specific to a geological period or faunal stage. The best index fossils are easy to identify at the species level and have wide distribution within a particular horizon, which means the likelihood of recognizing one in exposures of differing ages is low. The other answers are: a) paratype = a specimen that helps define the scientific name of a species and other taxon but which is not the holotype specimen. Paratypes may be listed as representative specimens in the original description (but see "d" for the difference between paratype and isotype). c) There's no specific term for this kind of fossil, though an index fossil could potentially occur only just before or just after an extinction event and, in that case, could help pinpoint the event. When Walter Alvarez studied the K-Pg (K-T) boundary in Italy, the number, species diversity, and size of microscopic foraminifera above and below the boundary were a fundamental clue that something dramatic had happened. d) holotype = the one specimen designated by an author as the basis for a taxon (usually a genus or species; but holotypes can exist for families and other taxonomic groups); if more than one specimen is used as the basis for erecting a new group or species, however, those duplicate specimens are called isotypes (this term is used only in botany, however).

She Sells Seashells Down by the Seashore

Born in 1799, she was an "amateur" fossil collector and paleontologist whose finds in and around southwest England contributed to important changes in scientific thinking about prehistoric life and the Earth's geological history. Though she was well-known in the scientific community of her day, she was not allowed to join the Geological Society of London because of her gender, was frequently robbed of credit for her discoveries, and remained poor her entire life. An 1823 article in the *British Mirror* said of her: "This persevering female has for years gone daily in search of fossil remains of importance at every tide, for many miles under the hanging cliffs, whose fallen masses are her immediate object, as they alone contain these valuable relics of a former world, which must be snatched at the moment of their fall, at the continual risk of being crushed by the half suspended fragments they leave behind, or be left to be destroyed by the returning tide—to her exertions we owe nearly all the fine specimens of Ichthyosauri of the great collections." In popular culture, she is said to have inspired the tongue-twister, "She sells seashells down by the seashore." What was her name and where was she born?

A: Mary Anning, born 21 May 1799 in Lyme Regis, Dorset, England.

20

Who Said It? (Part 1)

Name the paleontologist, geologist, or naturalist responsible for the following quotes.

1. "In many ways, we humans are the fish equivalent of a hot-rod Beetle. Take the body plan of a fish, dress it up to be a mammal, then tweak and twist that mammal until it walks on two legs, talks, thinks, and has superfine control of its fingers—and you have a recipe for problems. We can dress up a fish only so much without paying a price. In a perfectly designed world—one with no history—we would not have to suffer everything from hemorrhoids to cancer."

2. "The Ediacaran fossils disappeared right before the Cambrian Explosion, and it's an area of active scientific debate as to why they disappeared.... Some scientists think they went extinct due to changes in climate and environment. Other scientists think they were out-competed by the newly evolved animals, and still other scientists think they survived into the Cambrian but just weren't preserved in the rocks.... If [Ediacaran organisms] really were animals, then you and I might be considered the descendants of the Ediacarans. But there's also a good chance, even if they were animals, that the lineage that encompasses the Ediacaran fauna went extinct, so there might not be any living representatives of the Ediacaran fauna today."

3. "Of what use are the great number of petrifactions, of different species, shape and form which are dug up by naturalists? Perhaps the collection of such specimens is sheer vanity and inquisitiveness. I do not presume to say; but we find in our mountains the rarest animals, shells, mussels, and corals embalmed in stone, as it were, living specimens of which are now being sought in vain throughout Europe. These stones alone whisper in the midst of general silence."

4. "Many people are surprised when they learn that I study fossilized feces. I learned the informative value of scat during my tenure as a naturalist for the National Park Service. Park visitors on my guided nature hikes were bemused at my delight in finding animal feces, until I explained that such calling cards can provide important information about an animal's feeding habits and distribution. Unfortunately, fossil feces, known as coprolites, are more enigmatic than fresh fecal material."

A large, Miocene-era coprolite (just over 9" long and weighing in at about three pounds). Beaufort County, South Carolina. Photo: The Poozeum. Licensed under CC BY 4.0.

Brachiopods or Bivalves?

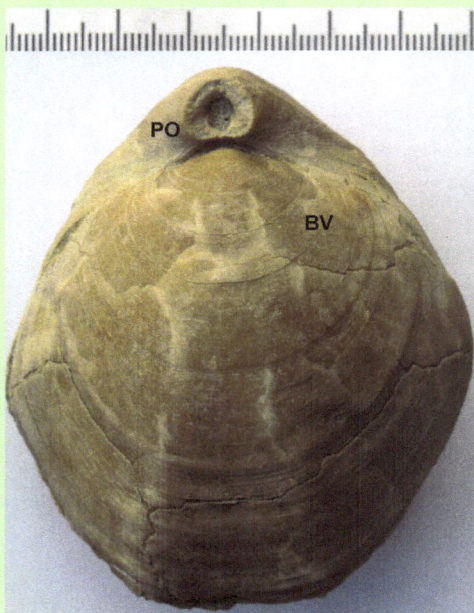

Collectors frequently misidentify bivalves as brachiopods and vice-versa. As a taxonomic matter, bivalves belong to the Phylum Mollusca, while brachiopods constitute their own phylum: Brachiopoda. In cladistics, the Phylum Bra-

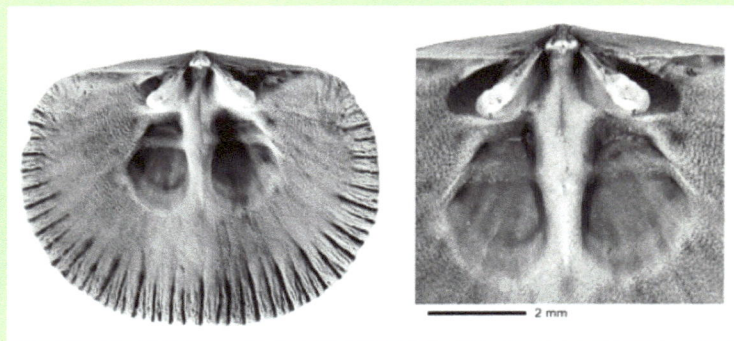

chiopoda occurs within the Clade Lophophorata along with Bryozoa and a small group of filter-feeding marine worms.

The internal anatomy of Brachiopod and Bivalve animals is dramatically different, but only the shells are preserved as fossils. When the interior of valves is well-preserved and visible, though, scars of muscles and other structures can also aid in identification.

Externally, there are several features that can help distinguish between the two groups, and mastery of a few facts will make it easy to tell one group from the other (see the key for specific information).

How well can you distinguish between bivalves and brachiopods using these features: *symmetry; pedicle opening (PO) or umbo (U); internal shell structures, including muscle scars, teeth, and the "pallial line" (PL); whether the valves are open or closed;* and *age*? (Hint: Start by using the key to identify these structures in the images shown here.)

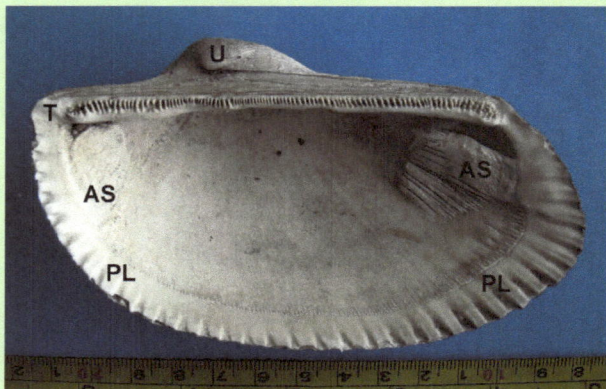

Brachiopods or Bivalves?

1. **Symmetry**: Generally speaking, bivalves have valves of the same size, and the two valves are symmetrical *between* them. In other words, if you pass a piece of paper between the valves, each valve will be a mirror image of the other. This is virtually always true of bivalves, though there are exceptions (rudists and oysters, for example). Brachiopods are also symmetrical, but the two valves are generally different in size and shape. The symmetry of brachiopods, then, is *perpendicular* to the hinge line. A line that passed from hinge line to margin at the midline of the shell would divide a brachiopod into two equal halves.

2. **Pedicle opening/umbo:** Bivalves generally have an obvious beak or "umbo" that curves inward toward the hinge line in both valves. What is never present in a bivalve, however, is a pedicle opening. Most brachiopods were stalked, and the stalk extended from an opening in the pedicle valve. This opening is especially pronounced in terebratulids.

3. **Internal shell structure:** On the interior of a single valve of any bivalve, look for the "pallial line." This is a demarcation within the shell, generally more-or-less parallel to the outer margin, which shows where muscles were attached in life. Roundish scars at either end of the pallial line sometimes also show where adductor muscles were attached. Neither of these features is present in brachiopods. What may be visible inside a brachiopod shell, however, is a looped or spiral structure in the brachial valve called a brachidium, which supported the brachiopod's feeding apparatus. Similarly, a line of many small teeth can often be seen along the hinge line in bivalves, but articulated brachiopods only sport two teeth in the pedicle valve, which correspond to two sockets in the brachial valve.

4. **Valves open vs. closed:** This is not a definitive feature, but both brachiopods and bivalves have muscles that keep their shells closed. Bivalves, however, also have a ligament that tends to force the valves open when the animal dies. With brachiopods, then, there is a somewhat greater likelihood of finding closed specimens. Obviously, much depends upon conditions around the time of death and before fossilization.

5. **Age:** Knowing the age of the exposure in which a specimen is found can also provide important clues. Though brachiopods dominated Paleozoic near-shore environments, a large number of brachiopod families never made it beyond the Permian. (Only about a hundred brachiopod species survive today, limited largely to deep seas and to three main groups vs. some 5,000 species that thrived during the Paleozoic). While a few Cretaceous brachiopods existed, for example, a two-valved fossil found in Cretaceous strata is much more likely to be a bivalve. Again, this is not definitive proof, but it is an important piece of information that aids in identification.

Key to images. Facing page, top left: The brachiopod Terebratula *sp. from the Pliocene of Italy. Note the large, prominent pedicle opening (PO) and the much different shape of the brachial valve (BV). A line drawn straight down from the middle of the pedicle opening would create two symmetrical halves. Bottom left: The bivalve* Anadara floridana *from the Florida Plio-Pleistocene. AS: adductor-muscle scar; PL: pallial line; T: teeth; U: umbo. Scale in centimeters. Photos © Wendell Ricketts. Middle right: The interior of the brachiopod* Cincinnetina multisecta *(Meek, 1873), showing cardinalia and adductor scars; from Jin, J. (2012).* Cincinnetina, a New Late Ordovician Dalmanellid Brachiopod from the Cincinnati Type Area, USA: Implications for the Evolution and Palaeogeography of the Epicontinental Fauna of Laurentia. *Palaeontology, 55(1), 205–228.*

Paleo Crossword II
... a challenging puzzle

Across

2. For Bryozoa, among other fossils, this is the section you often need for identification.
3. "Darwin's bulldog."
4. A fossil that identifies and dates the strata in which it is typically found or one of your fingers.
8. The end of the forelimb of a dinosaur is technically called this.
10. With 23 Down, an example of 7 Down.
11. They were originally named *Clupea*, but they're still the most common fish fossil in Wyoming's Green River formation.
12. Formerly, it was *Dynamosaurus imperiosus*.
14. This bony ring is a support around the eyes of several groups of vertebrates but may be particularly striking in ichthyosaurs.
15. Here's where lacustrine sediments are formed.
16. "The carpenter's daughter has won a name for herself, and deserved to win it." (Charles Dickens).
17. Tooth shells or tusk shells, some species were strung as ornaments or money by indigenous people.
18. The Osteichthyes have these.
19. This extinct Permian cartilaginous fish had a jaw that looked like a circular saw.
21. Formation in British Columbia, Canada, famous for soft-bodied fossils: the _____ Shale.
24. Some believe these "traps" may have contributed to the end-Cretaceous extinction.
26. When you're convinced it's a fossil but it's not.

Down

1. The shortest scientific dinosaur name (genus & species).
3. An ungulate gets around on these.
5. "Darwin's' pit bull".
6. These extinct Jurassic/Cretaceous bivalves may have had bad manners.
7. Fossil deposits that are extremely rich or where fossils are exceptionally well-preserved.
9. Phylogenetic systematics.
13. The study of the death, burial, and fossilization process of organisms.
18. Discovered in Alaska, this steppe mammoth has a colorful name.
20. If it weren't for this organic material, trilobites and insects would rarely have become fossils.
22. We're living in it.
23. With 10 Across, an example of 7 Down.
25. Paleontologist most associated with punctuated equilibrium.

25

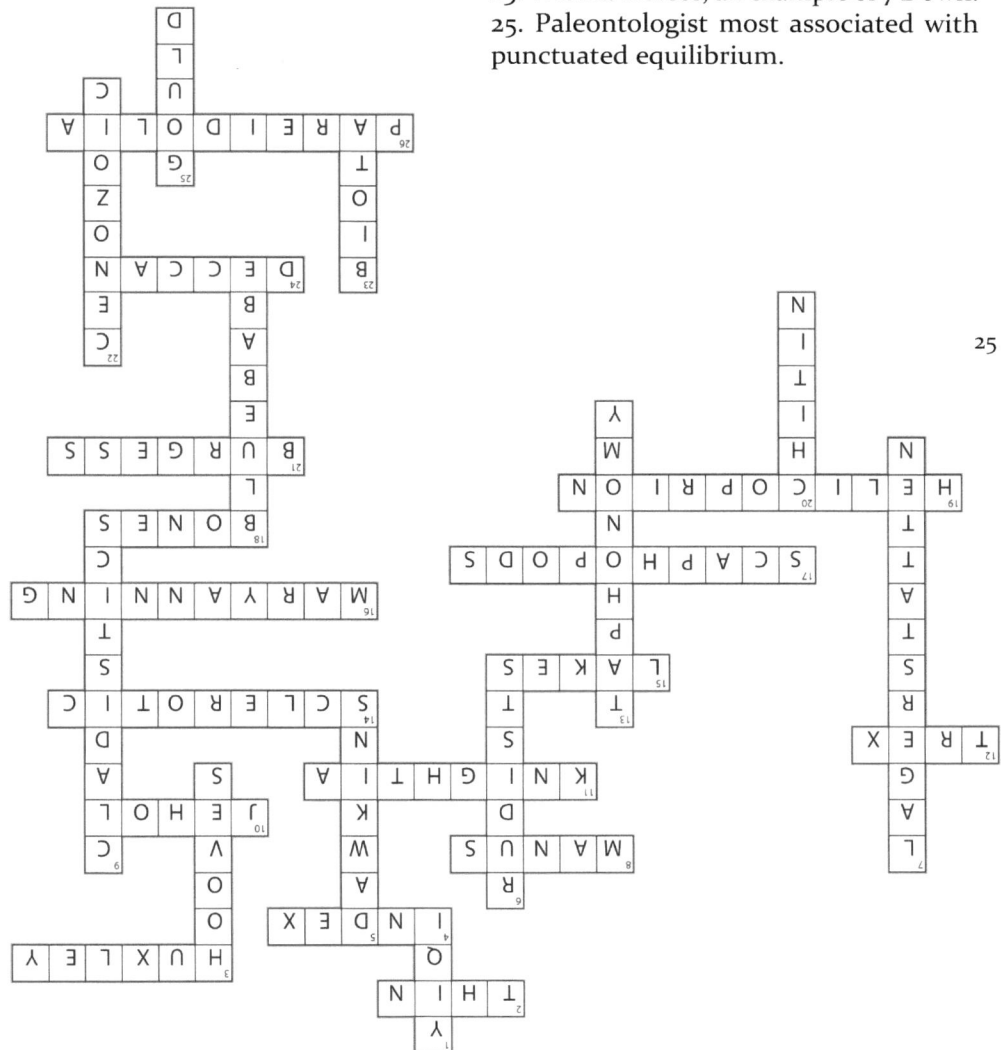

The drawing is by Everard Home, a British surgeon (1756-1832), from his paper, "Some Account of the Fossil Remains of an Animal More Nearly Allied to Fishes than any Other Classes of Animals," read before the Royal Society of London on June 23, 1814, and later published in *Philosophical Transactions of the Royal Society*, *104* (1814). It depicts part of the skeletal remains of the first ichthyosaur, named *Temnodontosaurus platyodon*. According to contemporary accounts, Joseph Anning found the skull in 1811, and Mary discovered the ribs and other parts of the skeleton the following season (she was 13 at the time). Interestingly, Neither Mary Anning nor her brother is mentioned in Home's paper, though he did record the name of William Bullock, the director of the Museum of Natural History in London, where the fossils were housed, and of Henry Host Henley, the owner of the land upon which the specimens were found.

Here are three hints: The image was not drawn by the person(s) who found this fossil, the image was first published in a scientific paper in 1814 (though the specimen had been discovered a few years earlier), and the scientific name the animal was originally assigned to was *Temnodontosaurus platyodon* (the "broad, cutting-tooth lizard").

What Is It? A Discovery More than 200 Years Old

Lagerstätten

A Lagerstätte (plural: Lagerstätten) is a fossiliferous deposit that yields specimens of animals or plants that are exceptional for their abundance and variety or because fossils are preserved with an extraordinary degree of detail, sometimes including soft tissues. Here's a list of a few of the world's Lagerstätten. Where is each one found, and what makes it famous?

Ashfall Fossil Beds

Burgess Shale

Ediacara Hills

Gogo Formation

Hunsrück Slate

Kaili Biota

La Brea Tar Pits

Sahel Aalma, Hjoula, and Haqel

Santana Formation

Ashfall Fossil Beds: Ash from a Yellowstone hotspot eruption 10-12 million years ago created these fossilized bone beds in northeastern Nebraska from which well-preserved Middle Miocene mammals (as well as some birds and reptiles) are known.

Burgess Shale: Located in the Canadian Rockies of British Columbia, Canada, the Burgess Shale was discovered in 1909 and continues to yield remarkable soft-bodied fossils dated to the Middle Cambrian.

Ediacara Hills: This range of low hills in the northern Flinders Ranges in South Australia contains the splendidly preserved fossils of strange and beautiful pre-Cambrian life forms.

Gogo Formation: Found in the Kimberley region of northwestern Australia, the Gogo preserves an exceptional Devonian reef community.

Hunsrück Slate: A Lower Devonian unit in the Hunsrück and Taunus regions of Germany famous for a highly diverse fossil fauna including brachiopods, corals, trilobites, plants, sponges, cephalopods, cnidarians, gastropods, and crinoids, to name a few.

Kaili Biota: A Lower to Middle Cambrian formation deposited in what is now part of the Guizhou Province in southwest China. Some Kaili species are also known from the Burgess Shale.

La Brea Tar Pits: Natural asphalt deposits in what is now the Hancock Park area of urban Los Angeles created traps for Pleistocene animals, including saber-tooth tigers, dire wolves, bears, sloth, and many others.

Sahel Aalma, Hjoula, and Haqel: These three distinct sites of largely Cretaceous deposits near the east coast of Lebanon have been known for their fossils since at least the 13th century and have yielded exquisitely preserved fish and lobsters as well as such unusual finds as snakes, birds, and octopods.

Santana Formation: Found in northeastern Brazil's Araripe Basin, the Santana preserves a shallow, Early to Late Cretaceous inland sea; its fossils include fish, numerous insects, plants, and dinosaur remains.

Who Said It? (Part 2)

"The skull before us belonged to some of the lower order of animals the teeth are very insignificant the power of the jaws trifling, and altogether it seems wonderful how the creature could have procured food."

Awful Changes.

Man found only in a fossil state. —— Reappearance of Ichthyosauri.

"A change came o'er the spirit of my dream". Byron.

A Lecture. — "You will at once perceive," continued Professor Ichthyosaurus, "that the skull before us belonged to some of the lower order of animals the teeth are very insignificant the power of the jaws trifling, and altogether it seems wonderful how the creature could have procured food."

The well-known caricature entitled *Awful Changes* was drawn by English geologist Sir Henry Thomas de la Beche (10 February 1796–13 April 1855) and was first published in 1830. The vignette depicts "Professor Ichthyosaurus" with a human skull, lecturing to other Mesozoic marine reptiles. The complete legend reads:

Man found only in a fossil state — Reappearance of Icthyosauri
"A change came o'er the spirit of my dream." Byron

A Lecture, — "You will at once perceive," continued Professor Ichthyosaurus, "that the skull before us belonged to some of the lower order of animals the teeth are very insignificant the power of the jaws trifling, and altogether it seems wonderful how the creature could have procured food."

De la Beche also painted the watercolor, *Duria Antiquior* ("A More Ancient Dorset"), in the same year (this page). He based his images on fossils found by Mary Anning and created the first representation of a scene from deep time based on fossil evidence. Because Anning had fallen on economic hard times, de la Beche had lithographs made of his painting that could be sold to benefit her. (Images: public domain.)

1

2

Little Trace Fossil, Who Made Thee?

Ichnology is the study of trace fossils—that is, the marks, tracks, burrows, and other hints of what long ago creatures or plants did (or had done to them). Ichnofossils may be rare or common at a site. In some cases, such as, for example, the famous Cambrian Blackberry Hill locality in Wisconsin, they may be the only type of fossil that can be found. In the case of some soft-bodied organisms (jellyfish or worms, for example), ichnofossils may be all that remains to be studied. Ichnofossils are typically given Latin names (genus and species) based on their form rather than on the species that made them, which is generally not known for certain. On these pages are examples of many kinds of ichnofossils. Use your imagination to see if you can guess what kind of organism made them (and why and how).

3

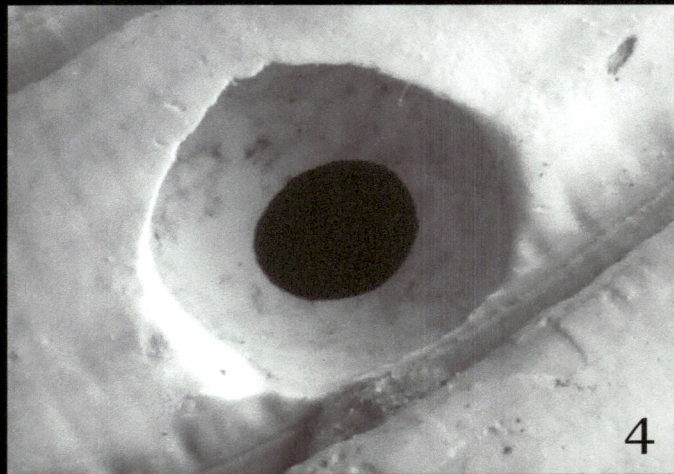

4

5

1.0 cm

6

2 cm

7

1.0 cm

8

9

10

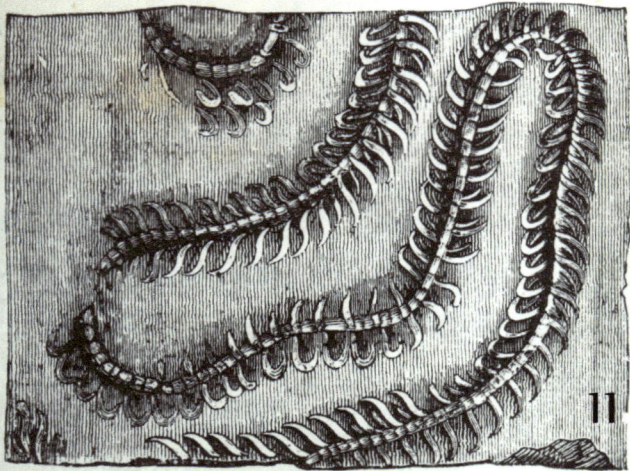

Ichnofossils

1. *Diplocraterion* sp. U. Ordovician, Kentucky. The U-shaped burrows of some type of suspension feeder. Candidates include crustaceans and worms, though even bivalves have also been hypothesized. The fossil is apparently a transverse section through the burrow with the wider indentations at each end of a horizontal furrow representing apertures.

2. *Rusophycus carleyi* (J. F. James). U. Ordovician, Ohio. Interpreted as the resting traces of trilobites. This image comes from Richard G. Osgood, Jr.'s *Trace Fossils of the Cincinnati Area*, a fascinating 1970 treatise that is available for free download. Osgood identified this specimen as the resting trace of the trilobite *Isotelus*.

3. ?*Grallator* sp. La Rioja Province, Spain. Dinosaur tracks are perhaps one of the most recognizable ichnofossils. A three-toed tetrapod left this footprint during the Jurassic or Cretaceous.

4. *Sedilichnus* sp., Plio-Pleistocene, Florida. This image shows a closeup of a drill hole in the bivalve, *Lirophora*, believed to represent predation by a carnivorous gastropod. © Wendell Ricketts; all rights reserved.

5. *Lockeia* sp. U. Devonian Chagrin Shale, Ohio. Believed to be the burrows of bivalves. Photo by Mark A. Wilson. Licensed under CC BY 3.0.

6. *Polykampton guberanum* Uchman, Wetzel & Rattazzi, 2019. Switzerland. These structures are interpreted as the dwellings of worm-like organisms that lived on flat, silted areas in the deep ocean. Image from Uchman, Wetzel & Rattazzi (2019).

7. *Entobia* sp. Recent, North Carolina. The numerous pits and holes on the shell of the bivalve *Mercenaria mercenaria* are believed to have been made by *Cliona*, a species of sponge. Photo by Mark A. Wilson. Licensed under CC BY 3.0.

8. *Petroxestes pera* Wilson & Palmer, 1988. U. Ordovician, Ohio. These slit-shaped structures were made by bivalves and, in this case, the associated species has been identified—*Corallidomus scobina*. Photo by James St. John. Licensed under CC BY 2.0.

9. *Asteriacites lumbricalis* Wilson & Rigby, 2000. Middle Jurassic, Utah. "Starfish" is an easy guess here—and it turns out to be right. An ophiuroid starfish settled down into the mud, leaving a five-rayed indentation that later filled with sediment and was preserved. Photo by Parent Géry. Licensed under CC BY 3.0.

10. *Chondrites* sp. U. Ordovician, Indiana. Unlike most of the other ichnofossils here, this one is interpreted as a feeding trace. The organism that made it tunneled beneath the surface of soft silt, making a tree-like pattern as it moved to the left and to the right in search of food. © Tim Volas; used by permission.

11. *Nereites cumbriensis*. From *La Terre avant le Déluge (The Earth before the Flood)* by Louis Figuier, 1863. Identified by Figuier as "a kind of annelid" from the Upper Silurian, adding that graptolites could be placed in the same group as well. (Graptolites are now believed to be a kind of colonial, filter-feeding animal rather than an annelid worm.) Image: Public domain.

12. *Protichnites eremita*, holotype. Cambrian, Blackberry Hill, Wisconsin. An example of the many forms of ichnofossils for which the Blackberry Hill site is considered a Konservat-Lagerstätte. This trackway, whose name means "first footprints," was likely formed on a sandy surface covered by a microbial mat (layered structures consisting of sediment bound by bacteria and the organism's secretions). From Kenneth Gass's *Solving the Mystery of the First Animals on Land: The Fossils of Blackberry Hill* (Siri Scientific Press). Photo by Susan Butts, courtesy of the Division of Invertebrate Paleontology, Peabody Museum of Natural History, Yale University.

13. *Daemonelix* sp. Middle Miocene, Nebraska. One of the most famous ichnofossils, *Daemonelix* mystified scientists and lay people for years. Commonly called "Devil's corkscrews," the large spiral burrows can be as much as two meters tall. In 1891, their discoverer, the paleontologist E. H. Barbour, interpreted them as a giant freshwater sponge, though he later decided they were "calcified plant forms." The theory that they represented rodent burrows dates to as early as 1895, but confirmation didn't come until the late 1970s, when a fossilized beaver, *Palaeocastor*, was discovered in one of them. Historical photo dating to late 19th or early 20th century; courtesy of the Agate Fossil Beds National Monument.

Who Said It? (Part 3)

Name the paleontologist, geologist, or naturalist responsible for the following quotes.

1. "If Mr. Hawkins has set the last specimen ... as it lay in the cliff, it will be a most magnificent specimen, but he is such an enthusiast that he makes things as he imagines they ought to be; and not as they are really found.... I would not have trusted [the specimen] to his making up[;] though very much broken, it might be made a splendid thing without any addition."

2. "The story of the Burgess Shale is also fascinating in human terms. The fauna was discovered in 1909 by America's greatest paleontologist ... Charles Doolittle Walcott, secretary ... of the Smithsonian Institution. Walcott proceeded to misinterpret these fossils in a comprehensive and thoroughly consistent manner arising from his conventional view of life: In short, he shoehorned every last Burgess animal into a modern group, viewing the fauna ... as a set of primitive or ancestral versions of later, improved forms. Walcott's work was not consistently challenged for more than fifty years."

3. "There comes a point, however, in studying dinosaurs ... when fossil bones can yield no more. The search for more information about how dinosaurs evolved has to shift to the genes of living creatures.... [W]e know that the history of evolution is written in the genes of modern dinosaurs, the birds, and how and why we can take a chicken egg that might have become part of an omelette or an Egg McMuffin, and convince it to turn into the kind of dinosaur we all recognize."

4. "Just above eye level, about eight feet high, there it was: three large dinosaur vertebrae and a femur weathering out of the cliff. It was so exciting because they were very large and because of the shape. The carnivores like *T. rex* had vertebrae [that were concave] from the disk.... The herbivores—the *Triceratops* or duck-bills, [which] is what you almost always find—have very straight vertebrae. So I knew it was a carnivore. I knew it was really big. And therefore I felt it must be *T. rex*, but it [couldn't] be *T. rex* because you don't find *T. rex*."

5. "It has been long and generally asserted ... that the lower animals were first introduced upon our globe, and formed along the population of the earliest periods in past time; that Polypi existed before Mollusks; these before Articulata, and that Vertebrata were the last to make their appearance. But the discoveries in fossil Ichthyology, which it has been my good fortune to describe in my researches upon fossil fishes, have shown that vertebrated animals, fishes, have existed in the oldest epochs...."

6. "Fossil Shells had long been known amongst the curious, collected with care, and preserved in their cabinets, along with other rarities of nature, without any apparent use. That to which I have applied them is new, and my attention was first drawn to them, by a previous discovery of regularity in the direction and dip of the various Strata in the hills around Bath ... which led me to the discovery of organic remains peculiar to each Stratum."

7. "[The] phone call I had gotten about giant raptors that morning ... was from one of the special-effects artists working in the *Jurassic Park* skunk works, the studio where the movie monsters for Spielberg's film were being fabricated in hush-hush conditions. The artists were suffering secret anxiety about what was to become the star of the movie—a raptor species of a size that had never been documented by a real fossil."

8. "Great white sharks, big storms - somehow, I think we like to be put in our place by awesome things. Dinosaurs do that."

1. Mary Anning, in 1832, revealing her concerns about how Thomas Hawkins (1810-1889), the English fossil dealer, would treat an ichthyosaur specimen she had found and which he had excavated from the cliffs at Lyme Regis. As it turns out, she was right to worry. Hawkins cast a great many missing or broken pieces of the specimen out of plaster, which he painted to resemble bone, and attempted to pass the specimen off as complete and original when he donated it to the British Museum. Quoted in *The Dragon Seekers: How an Extraordinary Circle of Fossilists Discovered the Dinosaurs and Paved the Way for Darwin* by Christopher McGowan. *New York: Basic Books, 2009:* 131-132.

2. Stephen Jay Gould, *Wonderful Life: The Burgess Shale and the Nature of History* (1989): 24.

3. Jack Horner in his and James Gorman's *How to Build a Dinosaur: The New Science of Reverse Evolution* (New York: Penguin, 2009: 14).

4. Sue Hendrickson, the discoverer in 1990 of "Sue the *T. rex*," the largest and best preserved *Tyrannosaurus rex* specimen ever found, quoted in *Tyrannosaurus Sue* by Steve Fiffer (W. H. Freeman & Co., 2000).

5. Louis Agassiz, *Twelve Lectures on Comparative Embryology*, 1849, Lecture IV, p. 27.

6. William Smith, *Strata Identified by Organized Fossils Containing Prints on Colored Paper of the Characteristic Specimens of Each Stratum* (London: W. Arding, 1816). Smith (1769-1839) was the English geologist who created the first nationwide geological map. His work was largely ignored during most of his life, but he ultimately became known as the "Father of English Geology."

7. Robert T. Bakker, *Raptor Red* (1995): 3.

8. Bonus Sue Hendrickson, the discoverer of "Sue the T. Rex," from an interview with the *Honolulu Advertiser,* July 9, 2000.

A Tiny Bit of Information: Microfossils

What kind of fossils can be called "microfossils" depends a lot on whom you're asking. The usual definition is "a fossil or fossil fragment that can be seen only with a microscope," which renders the general idea but leaves out a lot of possibilities. Many microfossils can actually be seen with the naked eye, but not in any detail—that is, they look like tiny grains of sand or fragments of rock. Magnification would be needed to see useful detail or make identifications. In other cases, fossils that are merely small (juvenile mollusks or tiny shark teeth, for example) are sometimes called "microfossils" for convenience, even though they can be seen well enough for identification purposes either without help or with the use of a magnifying loupe or visor. (Some collectors simply call these fossils "smalls" because there's not really a better word for them!) Other fossils are literally microscopic: cells of algae or most pollen grains, for example, though clumps or clusters of them may be visible without help. Super-tiny fossils are usually called "nannofossils."

For most purposes, paleontologists and collectors use what are called "binocular," "compound," "stereo," or "dissecting" microscopes, all of which are terms that describe an instrument with two sets of lenses that provide comparatively low magnification—from 20x to as much as 500x in the case of high-end research microscopes. Compound scopes are also often used in fossil preparation where delicate detail work is required.

High-magnification microscopes range from optical scopes (which use visible light and can magnify up to about 1500x) to electron microscopes (sometimes called SEMs or scanning electron microscopes) that can enlarge images up to an astonishing 160,000x. SEMs can also produce astonishingly detailed and beautiful images, even of everyday objects. *The Journal of Micropaleontology*, published since 1982, is entirely dedicated to microfossils, and all its articles are licensed via Creative Commons and are publicly available online.

Here are definitions of some relatively common terms in micropaleontology. What is the proper term for each one?

1. Tiny crustaceans sometimes called "seed shrimp." Some 70,000 species have been identified, the vast majority of which are extinct. Because their bodies are enclosed in hinged shells made of chitin or calcium carbonate, they are sometimes mistaken for bivalves. Most are marine, but they can live in freshwater as well as in soil itself. They are useful index fossils, meaning that they are distinct enough and widely distributed enough to be used to identify geological periods.

2. Small, skeletal elements of extinct and living sea sponges that help give the organism structural support and strength. They are made of calcium carbonate or silica and may occur in a huge array of shapes and sizes. As fossils, they date back to 665 million years ago.

3. Common and diverse fossils that are the components of the jaws of certain ringed (annelid) marine worms. Because they are made of chitin, they can be fossilized even when the bodies of their owners are generally (but not always) too soft to leave fossil traces. They have been found in the most ancient Paleozoic deposits but are most common in Ordovician, Silurian, and Devonian rocks.

4. Like the fossils above, these are the tooth-like elements of organisms that are not, in this case, worms, but are generally believed to be jawless vertebrates that resembled eels. Though tiny, they are often beautiful fossils whose shapes seem to suggest animals out of a fantasy novel. They existed in the world's oceans for more than 300 million years, from the Cambrian until their disappearance during the Jurassic. They, too, are useful index fossils.

5. One-celled microalgae found in soil and in fresh and salt water everywhere in the world. They are photosynthetic, like other plants and algae, and living species generate between 20% to 50% percent of the oxygen produced

on planet Earth each year. Their fantastic shapes may resemble ribbons, fans, zigzags, lanterns, jewels, or stars. Though individuals are tiny (from 2 to 200 micrometers), their silicon-based shells can create deposits that are as much as a half-mile deep on the ocean floor.

6. Minute calcareous platelets which, together, make up the shells of a kind of microscopic algae (called coccolithophores). Accumulations of these fossils may form chalk and, in fact, untold trillions of them make up the White Cliffs of Dover in England (along with other microfossils).

7. The name means "hole bearers," and these single-celled organisms commonly have an external shell (called a "test") that may be made of diverse materials and appear in a nearly infinite variety of forms and shapes.

8. BONUS. These two specific kinds of shelled, single-celled protozoans are members of the group identified above. Both can grow to be fairly large (up to two inches), so they may not all be microfossils, but many are quite a bit smaller. The name of the first kind means "little coins" and refers to their round shape. In cross-section, they often appeared to be coiled. They are particularly famous because the pyramids of Egypt were constructed using limestone, dating to the Eocene, that contains many of their fossils. The second kind is also named for its shape, which is typically (but not always) like a spindle or a grain of wheat—in fact, the genus name of one common species (*Triticites*) comes from the Latin word for wheat. They're especially common in Carboniferous and Permian rocks.

Note: µm is the symbol for microns. One micron (1 µm) is 1/1000th of a millimeter.

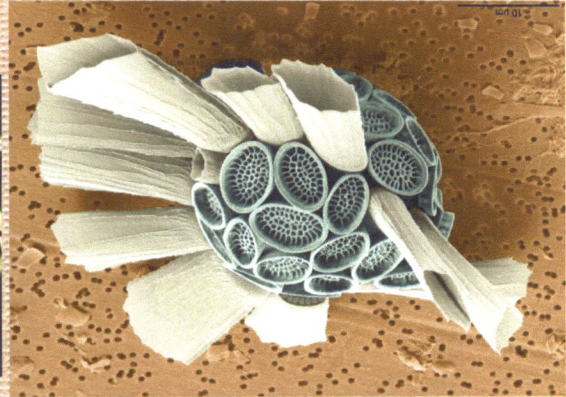

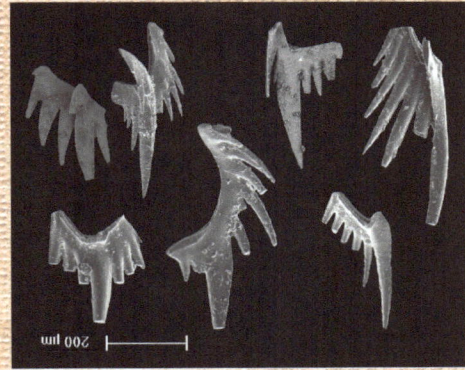

37

Counterclockwise from top left: 1. Conodont elements of *Neogondolella* from the Middle Triassic of British Columbia, Canada. From Martyn Lee Golding, "Reconstruction of the Multi-element Apparatus of *Neogondolella* ex gr. *regalis* Mosher, 1970 (Conodonta) from the Anisian (Middle Triassic) in British Columbia, Canada," Journal of Micropalaeontology, 2018. 2. SEM image of *Scyphosphaera porosa*, a rare deep-ocean coccolithophore (coccolith-bearing organism) from the South Atlantic. Image by Jeremy Young; © The Micropalaeontological Society; used by permission. 3. Strew of Early Miocene diatoms from St. Laurent la Vernède, France. © Stefano Barone (www.diatomshop.com).

1. Ostracods; 2. Spicules; 3. Scolecodonts; 4. Conodonts; 5. Diatoms; 6. Coccoliths; 7. Foraminifera; 8. Nummulites and fusulinids.

```
c s d z k x r j f t l e p t a e n a e n x t w p
o t b r h y n c h o t r e m a z s u f i i l c a
n r f l e x i c a l y m e n e a p a a d t c e d
s o e d o n d q c a t a z y g a m i c e e e o w
t p n e k v p i u e w d m m m y h e d a a t o x
e h g l y p t o r t h i s p v w u e e y h u d f
l o n g b a e u l d i p l e c t o r t h i s e d
l m t a m y e y p l a e s i o m y s n e l a a r
a e h y i n i c t c h e b e r t e l l a c u c y
r n m c i n c i n n e t i n a z y g o s p i r a
i a i a i u o n o j m r h i s c o b e c c u s i
a p n f j m g l u y a e a c y c l o n e m a r p
p l a t y s t r o p h i a u o a g f a y x j u r
h o l t e d a h l i n a z t r m h s z t e j a o
s o w e r b y e l l a i y y y i h v d g m a u t
r a f i n e s q u i n a m b e g n r v e e f q a
l e p t a e n a s e e m i r e k x u z w t g c r
p e t r o c r a n i a e n r o c v y s e p q p a
v i n l a n d o s t r o p h i a c a u r n e c e
b p t t n m j y d g g r e w i n g k i a k v v a
```

How Many of These Do You Know?*

Catazyga Ceraurinus Cincinnetina Constellaria Cyclonema Flexicalymene
Glyptorthis Grewingkia Hebertella Hiscobeccus Holtedahlina Leptaena Leptaena
Petrocrania Plaesiomys Platystrophia Plectorthis Protaraea Rafinesquina
Rhynchotrema Sowerbyella Strophomena Vinlandostrophia Zygospira

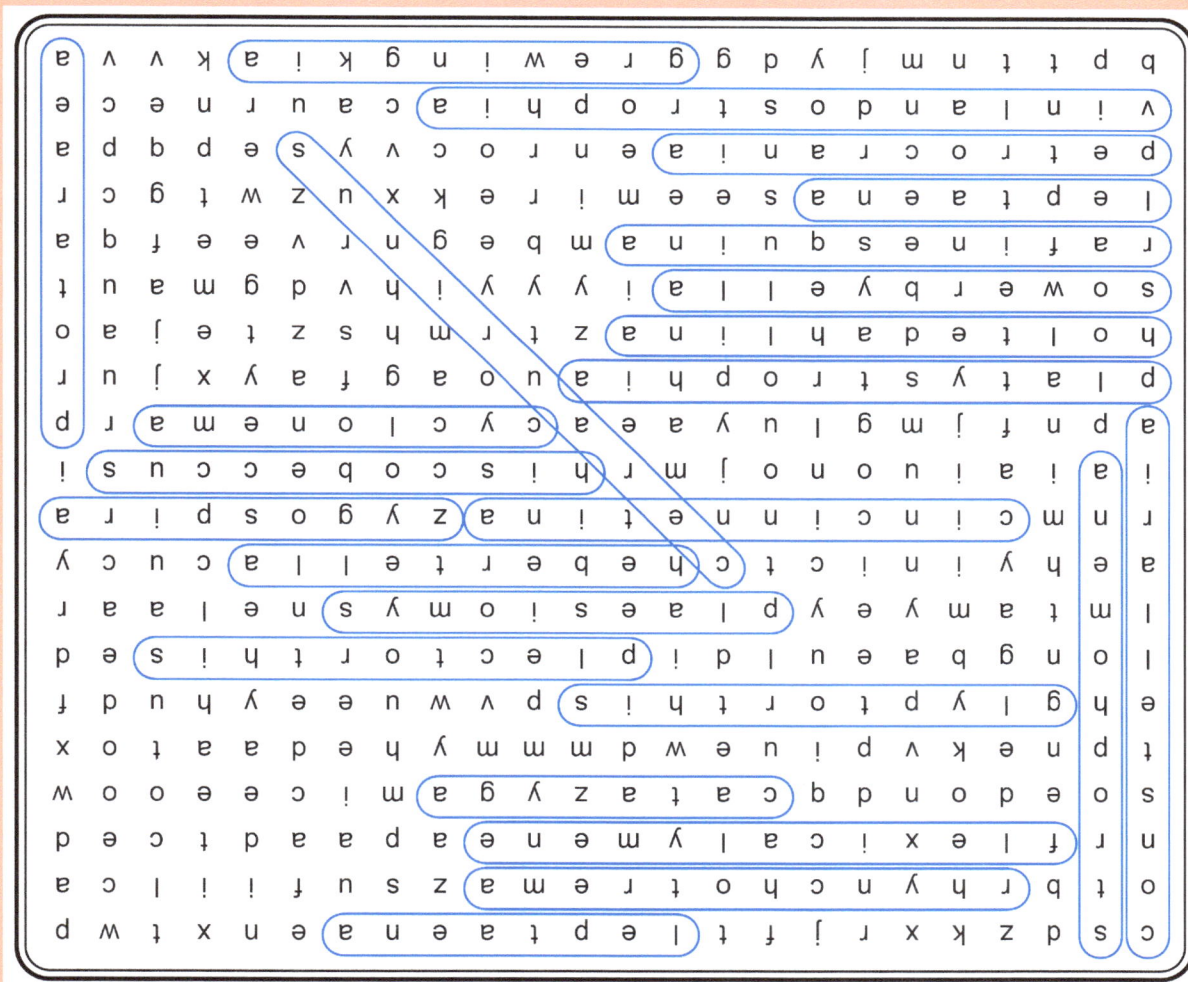

A couple of good places to see images of these fossils are "Cincinnati Fossils" on the UGA Stratigraphy Lab site (strata.uga.edu/cincy/fauna/fauna.html) and the Dry Dredgers "Fossil Photo Index" (drydredgers.org/thumb_index.htm).

Collectorese

Bugs, trikes, megs, hildis, and hemis ... the list seems endless.

There's something about fossil collecting that seems to bring out the desire to invent nicknames—and then to use them to play "I know something you don't know" with anyone who gets within spitting distance.

Let's not mince words: It's annoying. Please stop.

Granted, the things we're talking about are way, way deader than Latin, and sure: We know. No one wants to come off sounding like he's carrying actual scientific knowledge in his head, especially not these days.

Still, dropping cute little names into the conversation so other people will have to ask you what you mean is just sad. Honestly, speaking of nicknames, it could even be called a Richard move.

Now, you're welcome to get out among baseball fans and brag about your favorite player's awesome "romer" (because he had to "run home"), but don't blame us if they look at you funny. And no one's stopping you from going to your mechanic and saying you've got a problem with your sklugs, but you'll still be paying $95 an hour while she or he figures out what you're talking about.

Look, just use the right names for things. The time you save by skipping a few syllables doesn't add up to much. Science thanks you. Other collectors thank you. The fossils thank you.

Meanwhile, want to test your ability to decipher "Collectorese"? Here are some fairly common fossil nicknames. How many can you recognize—and do you know what they really mean? (No, I have no idea why so many of them have to do with shark teeth.)

ammos	flexis
angys	glypts
avits	hemis
barracudas	hildis
bennies	megs
brachs	pachys
bugs	priskies
cephs	rics
chubs	tikes
echies	trikes

ammos: Ammonites.

angy/angie: *Carcharocles angustidens*, an extinct shark.

avit: The Cretaceous crab, *Avitelmessus grapsoideus*. "Avit crab" and "avit nodule" are also used.

barracuda: Usually *Enchodus*, a ray-finned fish that spanned from the Cretacous to the Eocene. (It wasn't a barracuda.)

beni: *Parotodus benedenii*, an Otodontid shark with teeth especially prized by collectors.

brachs: Brachiopods.

bugs: An annoying term for trilobites which weren't "bugs" (members of Order Hemiptera) at all. In fairness, trilobite studies took off in the mid-18th century after Charles Lyttelton submitted a letter to the Royal Society of London in 1750 describing the "petrified insects" he had found, a species of *Calymene* that came to be widely known as the "Dudley Locust."

cephs: Cephalopods.

chub: *Carcharocles chubutensis*, a megatoothed Oligocene-Pliocene shark. It is considered a close relative of *O. megalodon*, but its classification is far from settled.

drum: Typically used to mean various fossil fish in the Order Pycnodontiformes or Family Pycnodontidae (but true drum fish belong to a different order and family).

echies: Echinoids.

flexis: *Flexicalymene*, a common Upper Ordovician trilobite and index fossil.

glypt: For *Glyptodon*, a genus of large Pleistocene armored mammal

hemi: An extinct species of shark from the Miocene, *Hemipristis serra*.

hildis: A fairly common Jurassic ammonite, *Hildoceras*.

megs: The so-called "megalodon" (which has a common but not scientific meaning). *C. megalodon*, though, is itself, a nickname. The genus was once either *Carcharodon* or *Carcharocles*, depending upon whether one favored the *Otodus*-origin or the *Isurus*-origin hypothesis. (Admittedly, both sound like the sound the cat makes right before heaving up a hairball.) In any case, it's now called *Otodus megalodon*. The first person who suggests "otes" is asking for it.

pachys: *Pachycephalosaurus*, a genus of "thick headed" dinosaurs from the Late Cretaceous.

priskies: *Priscacara*, an extinct Eocene perch, common in the Green River Formation.

ric: *Carcharocles auriculatus*, another large Otodontid Cretaceous shark, considered by some to belong to the genus *Otodus*.

tikes: *Ptychodus*, a Late Cretaceous shark with crushing plates for teeth.

trikes: A Cretaceous horned dinosaur, *Triceratops*.

BONUS 1: Trex, T-rex, trex, TRex, or, may the deities help us, "rexies." No. Just no. Use *T. rex* if you need an abbreviation.

BONUS 2: "Poop." If you are four years old, you have permission to use this word while discussing coprolites. Otherwise, say fossil dung or fossil feces. We know, the latter has four whole syllables, but we have faith in you.

1. *Homotelus*: A genus of trilobite, one species of which, *Homotelus bromidensis*, has yielded numerous specimens from the Middle Ordovician Bromide Formation in Oklahoma, where scores of complete trilobites have been found on slabs in densities as high as 170 individuals per square meter. Several hypotheses have been proposed to explain these concentrations.

2. Biloba: The species name of the *Ginkgo biloba* ("two lobes"), a tree native to China but now widespread. G. *biloba* appeared in the Middle Jurassic, and fossil and living G. *biloba* leaves still resemble each other very closely.

3. *Heterophrentis:* A Devonian solitary horn coral.

4. *Bison*: A common name for this animal but also part of a Latin binomial (*Bison bison*, meaning "wild ox") that indicates the American buffalo.

5. Bivalvia: A large class of mollusks that include the familiar clams, oysters, cockles, and similar species, most of which (but by no means all), are or were marine. The literal meaning is "two valves." The bivalves date to the earliest Paleozoic formations.

6. *Heterodiadema*: An extinct Cretaceous sea urchin known primarily from the Middle East.

A slab of Homotelus bromidensis *from Oklahoma. The forked structures visible on two individuals are hypostomes, a hard mouth part. Photo © James St. John. Licensed under CC BY 2.0.*

42

7. Binomial: Followed by "system" or "nomenclature," the lowest level in the ranked organization of biological life developed by Linnaeus in the 18th century. Literally: "two names" (specifically, genus and species). May refer to the system or to the name of an organism. For example, the "binomial name" for the housecat in "binomial nomenclature" is *Felix catus*.

8. Heteromorph: A group of extinct ammonites (Suborder Ancyloceratina) that first evolved during the Late Jurassic but didn't become common until the Cretaceous period, at which point they began to develop complex, beautiful, and fascinating shells. They're "different shape" ammonites because their shells are not regular spirals but rather may coil in several directions or planes.

Hetero, Homo, Bi

In zoology, geology, and paleontology, words that begin with "hetero," "homo," or "bi" are fairly common. To what do the words below refer?

1. Homotelus
2. Biloba
3. Heterophrentis
4. Bison
5. Bivalvia

6. Heterodiadema
7. Binomial
8. Heteromorph
9. Heterorhea
10. Heterostracan

11. Bicentenaria
12. Homo erectus
13. Homolidae
14. Heterodontosaurus
15. Homology

Left, top: *A rare fossil heteromorph ammonite,* Crioceratites nolani, *from the south of France. Licensed under CC BY 4.0.* Left, bottom: *A Ginkgo* biloba *leaf fossil from the Eocene Klondike Mountain Formation, Republic, Washington. Photo by Kevmin/Karl. Licensed under CC BY 2.0.* Bottom, right: *The jaws of* Heterodontosaurus tucki *from the Lower Jurassic of South Africa. Photo by Paul C. Sereno. Licensed under CC BY 3.0.*

---

9. *Heterorhea*: The "different rhea" is an extinct flightless, large, long-necked, and long-legged bird in the rhea family and is known only from the Late Pliocene Hermoso Formation in Argentina. The original fossil that was used to describe the genus (the holotype) has been lost.

10. Heterostracan: Literally "different scales" because of the arrangement of scales in varying patterns on their bodies. This extinct subclass of jawless vertebrate lived primarily in marine and estuary environments. The first species appeared during the Early Silurian and all, except for one group, became extinct by the start of the late Devonian. That last group died out in the end-Devonian extinction.

11. *Bicentenaria*: A genus of carnivorous, Late Cretaceous theropod that lived in what is now Argentina. It was named for the 200th anniversary of the 1810 May Revolution in Argentina, the first in a series of events that led to Argentina's War of Independence from Spain.

12. *Homo erectus*: *Homo* is the genus to which human beings and many of our ancestors belong. *H. erectus* appeared as many as two million years ago (*Homo sapiens*, in contrast, dates to only about 300,000 years ago), and this "upright man" is believed to have moved and walked essentially as modern humans do. The fossils of *H. erectus* were first unearthed in the late 1800s. Some twenty species of *Homo* (and numerous subspecies) have been proposed since the first half of the 19th century.

13. Homolidae: The Family Homolidae contains fourteen genera of mostly deep-water marine crabs. They are known commonly as carrier or porter crabs because living species carry objects like sponges, corals, or sea urchins on the back of their carapaces. Fossil specimens are relatively rare, but the group is known from as early as the Cretaceous.

14. *Heterodontosaurus*: An Early Jurassic dinosaur and the best-known member of the Family Heterodontosauridae, a group considered basal or "primitive" within the Order Ornithischia. A small dinosaur, its name means "different teeth" and refers to the fact that *Heterodontosaurus* had three different kinds: small, incisor-like teeth in the upper jaw; long, canine-like tusks; and cheek teeth like chisels.

15. Homology: A term that refers to the similarity between structures in organisms that belong to entirely different groups. For example, the five fingers in the hands of human beings are homologous with the feet of lizards, the paws of raccoons, and even the flippers of toothed whales—in fact, with all creatures in which this trait is shared because it developed from the same ancestral tetrapod structure.

Fossil Names: Greek & Latin Prefixes & Suffixes

The scientific vocabulary of biology and zoology—and, thus, of paleontology—is heavily influenced by words derived from Greek and neo-Latin. Match these relatively common prefixes and suffixes with their meanings.

1. a/an | **2.** archaeo | **3.** branch | **4.** cephalo | **5.** ceras | **6.** chondr | **7.** cteno | **8.** dactyl **9.** derm | **10.** dino/deino | **11.** echino | **12.** eo | **13.** gnathus/gnatha | **14.** gymn/gymno | **15.** hippo **16.** ichthyo | **17.** pachy | **18.** phor, phora | **19.** platy | **20.** pod/ped | **21.** ptero | **22.** saur/sauro **23.** smilo | **24.** sperma | **25.** stego | **26.** tetra | **27.** therium | **28.** vore/vorous

A. ancient
B. bearing, carrying
C. beast
D. cartilage
E. comb
F. early, primeval
G. eating, having a diet of
H. finger
I. fish
J. flat
K. foot
L. four
M. gills
N. head
O. horn
P. horse
Q. jaw
R. double-edged knife, scalpel
S. lizard (or reptile generally)
T. naked, bare
U. roof or covering
V. seed
W. skin
X. spiny
Y. terrible, frightening
Z. thick
AA. wing, winged
AB. without, lacking

44

1. AB | 2. A | 3. M | 4. N | 5. O | 6. D | 7. E | 8. H | 9. W | 10. Y | 11. X | 12. P | 13. Q
14. T | 15. P | 16. I | 17. Z | 18. B | 19. J | 20. K | 21. AA | 22. S | 23. R
24. V | 25. U | 26. L | 27. C | 28. G

Once you've confirmed your answers, you shouldn't have too much trouble getting at least a rough idea of the meaning of these terms:

- agnatha
- archaeopteryx
- branchiopod
- chondrichtyes
- ctenophora
- dinotherium
- echinoderm
- eohippus
- gymnosperm
- insectivorous
- pachycephalosaurus
- platyceras
- pterodactyl
- smilodon
- stegosaurus
- tetrapod

Two species of *Platyceras* from the Mississippian (Carboniferous) Burlington Limestone. From Charles R. Keyes, "On the Attachment of Platyceras to Palaeocrinoids, and its Effects in Modifying the Form of the Shell." *Proceedings of the American Philosophical Society*, Vol. 25, 1888.

agnatha: "without jaws." A group of jawless fish that date to the Cambrian; modern representatives include lampreys and hagfish.

Archaeopteryx: "ancient (or old) feather (or wing)." A genus of bird-like dinosaur that lived in the Late Jurassic, in what is now southern Germany, during a time when Europe was an archipelago of islands in a shallow warm tropical sea; it grew to about the size of a raven.

Branchiopod: "gill foot." These mostly small, fresh-water crustaceans (fairy shrimp, e.g.) feed on plankton and detritus and carry their gills on their legs.

Chondrichthyes: "cartilaginous fish." Fish without a bony skeleton. Sharks, skates, and rays are the most common living and fossil representatives.

Ctenophora: "comb-bearing." These marine invertebrates, commonly known as comb jellies, are named for the groups of cilia, or thread-like projections, that sometimes seem to move in waves over the animal's surface as it swims.

Deinotherium: "terrible beast." A large Miocene-Early Pleistocene elephant-like herbivore with tusks that grew from the lower jaw rather than the upper jaw as in modern elephants. Maybe not so "terrible" after all?

Echinoderm: "spiny skin." Members of the large phylum that includes sea urchins, star fish, crinoids, and others.

Eohippus: "ancient horse." The name pretty much says it all!

Gymnosperm: "naked seed." The seeds of this group of plants, whose members include pines, cycads, and *Gingko*, are not enclosed within a pod.

Insectivorous: "insect-eating." A huge group of animals and plants—ranging from anteaters to bats to spiders to Venus Fly Traps—subsist on insects, either exclusively or in part.

Pachycephalosaurus: "thick-headed lizard." A Late Cretaceous dinosaur with a massive dome on the top of its skull. The most popular theory about the dome's function is that pachycephalosaurs used it to fight each other.

Platyceras: "flat horn." A extinct marine gastropod, relatively common in the Paleozoic, whose shells roughly resemble a curved, somewhat flattened horn.

Pterodactyl: "wing finger" or "finger wing." Members of the Suborder Pterodactyloidea, one of the two suborders of pterosaurs, pterodactyls were flying reptiles (but not bird ancestors) that appeared during the Jurassic. Most pterodactyls lacked teeth.

Smilodon: "knife tooth." An extinct genus of saber-toothed feline that lived during the Miocene and Pliocene. The hundreds of specimens of *Smilodon* from the La Brea Tar Pits in Los Angeles, the largest collection in the world, have helped *Smilodon* become the best known example of the Machairodontinae or saber-toothed cats.

Stegosaurus: "roof lizard." Probably one of the most recognizable dinosaurs, *Stegosaurus* was an armored herbivore that appeared in the Late Jurassic. The arrangement of large spikes at the end of its tail came to be called the "thagomizer," a term coined as a joke by *Far Side* cartoonist Gary Larson in 1982.

Tetrapod: "four feet." The four-limbed vertebrates of this superclass include extinct and living amphibians, reptiles (including dinosaurs and therefore birds), and mammals—in other words, just about everything with a backbone.

The Bone Wars

Q: One of the greatest — and most destructive — rivalries among professional paleontologists in American history was between these two men, one from the Academy of Natural Sciences in Philadelphia and the other from the Peabody Museum of Natural History at Yale. In their rush to be the first to find and describe previously unknown "saurians" from the American west, they resorted to bribery, theft, bitter attacks on one another's reputations and character, accusations of plagiarism, destruction of specimens, and (allegedly) dynamiting of fossil locations so others could not collect there. Their decades-long feud came to be known as the "Bone Wars." Who were they, and where and when did they work?

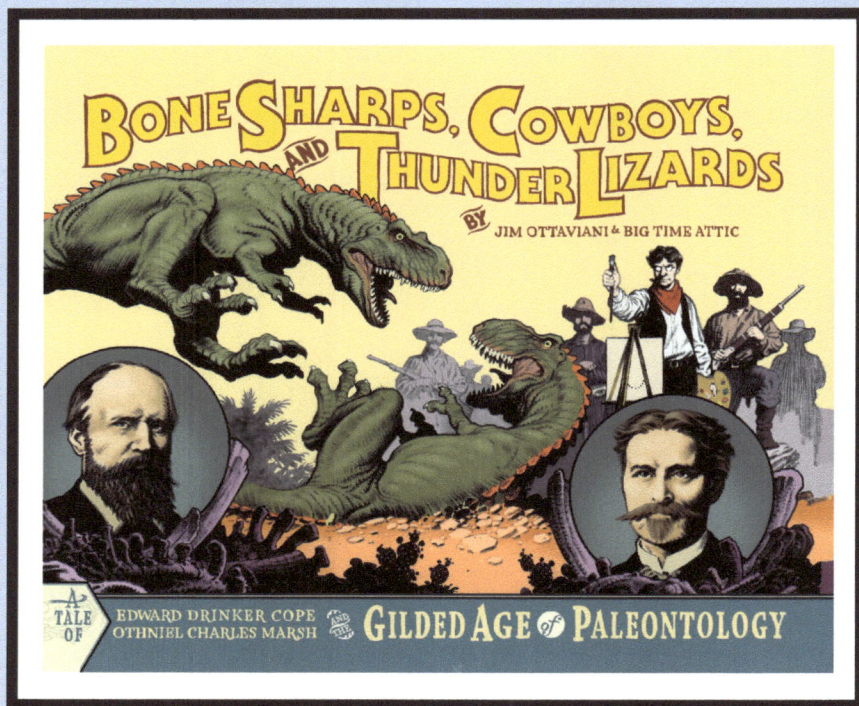

In text and drawings, Jim Ottaviani's *Bone Sharps* tells the story of "gilded age" fossil hunters and the American "Bone Wars."

A: Edward Drinker Cope (Academy of Natural Sciences) and Othniel Charles Marsh (Peabody Museum of Natural History). Most of their work was done in the bone beds of Colorado, Nebraska, and Wyoming between 1877 and 1892. When Cope died in 1897, the two men were still sworn enemies, but the decades of trying to outdo each other had left them both all but bankrupt. Michael Crichton's 1974 novel, *Dragon Teeth*, published posthumously in 2017, is a historical novel set during the Bone Wars and is based on the real life Marsh and Cope.

1. The geological periods: PreCambrian, Cambrian, Ordovician, Silurian, Devonian, Carboniferous, Permian, Triassic, Jurassic, Cretaceous, Paleocene, Eocene, Oligocene, Miocene, Pliocene, Pleistocene, Recent. For those who prefer "Holocene" to "Recent": Plentiful early oiling might prevent permanent hernias.

2. The five major mass-extinction events at the end of the Ordovician, Devonian, Permian, Triassic, and Cretaceous.

3. The periods of the Paleozoic: Cambrian, Ordovician, Silurian, Devonian, Mississippian, Pennsylvanian, Permian.

4. The periods of the Mesozoic: Triassic, Jurassic, Cretaceous.

5. The timeline of the Cenozoic: Paleocene, Eocene, Oligocene, Miocene, Pliocene, Pleistocene, Holocene. If you prefer "Recent" to "Holocene," replace "heartily" with "repeatedly." An alternative: "Put eggs on my plate, please Homer."

6. The major taxonomic ranks: Kingdom, Phylum, Class, Order, Family, Genus, Species. (The mnemonic leaves out the less commonly used "Domain," which occurs before "Kingdom" in the three-domain system.) There are dozens of alternatives, including "Koalas prefer chocolate or fruit, generally speaking," "Kids prefer cheese over fried green spinach," "Drunken kangaroos punch children on family game shows," "Keep pots clean or family gets sick," "Kings play chess on fine- grained sand," and "Dammit, kitty. Please come out for goodness sakes! (in the voice of Eric Cartman).

BONUS. The classes of Phylum Mollusca and subclasses of the Cephalopoda. Scaphopoda, Gastropoda, Caudofoveata, Solenogastres, Monoplacophora, Polyplacophora, Bivalvia, Cephalopoda, CAN (Subclasses Coleoidea, Ammonoidea, and Nautiloidea). Admittedly, this mnemonic is slightly outdated. It leaves out the Rostroconchia and Helcionelloida, two extinct classes included under the mollusks in comparatively more recent classification schemes. It also doesn't reflect current debates about Subclass Orthoceratoidea, which is considered by most workers to belong under Cephalopoda.

BONUS: Some grownups can't see magic ponies, but children CAN!

6. Keep parks clean or fires get started.

5. Pigeon egg omelets make people puke heartily.

4. The jumping cat.

3. Can Oscar see down my pants pocket?

2. Older deer play together cautiously.

1. Pregnant camels often sit down carefully. Perhaps their joints creak? Plentiful early oiling might prevent permanent rheumatism.

There's a lot of terminology in geology and paleontology, and students, professors, and researchers have created a lively assortment of mnemonic devices—ranging from the silly to the downright obscene—to help keep things straight. Here are a few of the cleaner ones. We give you the mnemonic; you guess what it's meant to bring to mind.

Paleomnemonics

Paleo Jeopardy!

Here are the answers. What's the question?

Be sure to fill in the blanks on the facing page with the entire question. Don't add spaces between words.

Example:

Answer: In 1981, one commercial collector estimated that, over the previous twenty years, he'd extracted and sold 1.5 million specimens of this common trilobite species from Utah's Wheeler Shale.

Question:

W	H	A	T	I	S	E	L	R	A	T	H	I	A	K	I	N	G	I	I

The Answers:

1. These curved, shelly plates are now understood to be part of the body of an ammonite, though they are almost always found separated from ammonite shells.

2. If you reorganize your "Cat Emails," you'll find the name of this genus of extinct Carboniferous horsetails.

3. *Pecten*, *Ostrea*, and *Conus*, for example.

4.   A lot of Pleistocene mammals got stuck in this pleonasm.

5. For paleontologists, it's better to get your fly caught in this than something caught in your fly.

6. This frequently found segment of a trilobite body is jokingly called a "trilobutt."

7. A jumping one of these helps you remember the periods of the Mesozoic.

8. Change one letter in the name of this geological epoch to get the name of the system that contains it.

9. You may have the right to bear these, but so do starfish.

10. *Dinohippus* was a terrible one of these.

11. The reason why the east coast of South America and the West coast of Africa seem to fit together like jigsaw puzzle pieces.

12. Birds do it, bees do it, maybe even *Oviraptor* did it.

13. You have one of these in your nose but it's also found in some corals.

14. This ultra-common Moroccan Ordovician trilobite is offered for sale by the hundreds at gem-and-mineral shows around the world.

15. *Mei long, Bolong, Zuolong,* and *Shaochilong* are all species of dinosaur known from China, but the suffix "long" means this rather than "lizard" or "reptile."

49

1.	W	H	A	T														
2.	W	H	A	T														
3.	W	H	A	T														
4.	W	H	A	T														
5.	W	H	A	T														
6.	W	H	A	T														
7.	W	H	A	T														
8.	W	H	A	T														
9.	W	H	A	T														
10.	W	H	A	T														
11.	W	H	A	T														
12.	W	H	A	T														
13.	W	H	A	T														
14.	W	H	A	T														
15.	W	H	A	T														

Unscramble the letters in the darker boxes to reveal the name of this important Lagerstätte for unusual early invertebrates.

———— ———— ———— ———— ———— ———— ———— ———— ———— ———— ———— ———— ———— ———— ————

1. What are aptychi?
2. What is *Calamites*?
3. What are mollusks?
4. What are The La Brea Tar Pits? (Literally, "the the tar pits")
5. What is amber?
6. What is a pygidium?
7. What is a cat? (*The Jumping Cat* = Triassic, Jurassic, Cretaceous)
8. What is the Paleocene? (Change one letter to get Paleogene.)
9. What are arms?
10. What is a horse? (The name literally means "terrible horse.")
11. What is Plate Tectonics Drift?
12. What is build nests?
13. What is a septum?
14. What is *Flexicalymene*?
15. What is dragon?

The letters **HUÉRTFAHSEBSSLG** spell out "The Burgess Shale."

Also from FourCats Press

A concise introduction to understanding what gives a photograph its power to move us and to make us think. The book guides the reader through basic elements of what constitutes a photograph and then presents a method of analysis to help viewers understand why a photo is compelling and how it conveys its message. Ideal for students of photography, for art lovers, and for photographers in search of a brief guide to discussing and understanding photographs and photography. ISBN: 978-1732044289 | 8.5" x 8.5" | 127 pages

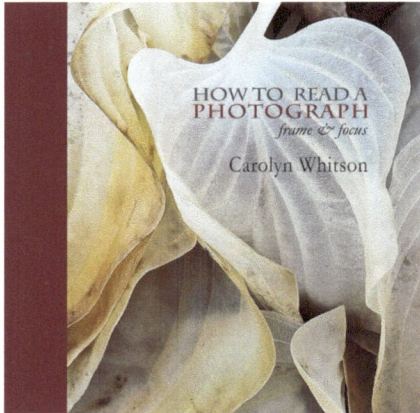

How do adults know when something is for boys and when it's for girls? Who tells them so? Where do they learn it? For eight-year-old Luca, it's a mystery, but if he can't convince his parents to give him the white ice skates he has his heart set on, Christmas is going to be ruined. Who does a child turn to when he can't even count on Santa Claus? [ISBN: 978-0989980029 | 5" x 8" | 66 pages]

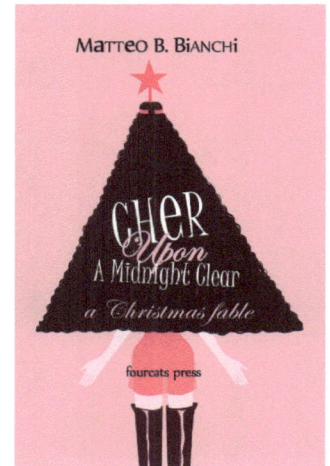

Fossil News: National and international news from the world of vertebrate and invertebrate paleontology, advice for amateur and para-professional fossil collectors, updates on important issues in paleontology, paleoart & photography, original book reviews, and many other features. Back issues from 2016-2022 available for order at a discount from FourCats Press (fossilnews.org/backissues).

November 12, 2003: A suicide attack on the Italian military base in Nasiriyah, Iraq, leaves 19 dead and scores wounded. Among the survivors is a young, brash, naïve Italian filmmaker, Aureliano Amadei. This is the story of a man who arrived in the midst of the terror, fire, and gunfire of a war that officially didn't exist ... and who came away both permanently changed and more determined than ever to tell the truth of what he experienced. [ISBN: 978-0989980005 | 5.5" x 8.5" | 188 pages]

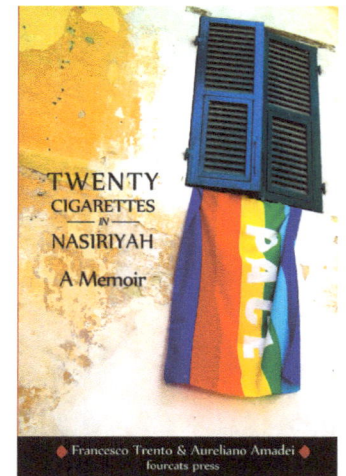

www.FourCatsPress.com

www.ingramcontent.com/pod-product-compliance
Lightning Source LLC
Chambersburg PA
CBHW060843200326
41521CB00003BB/162

*9 7 8 1 7 3 4 8 0 5 0 3 1 *